DUALITY OF VOICE

AN OUTLINE OF ORIGINAL RESEARCH

BY

EMIL SUTRO

1899

Theophania Publishing

INTRODUCTION

By the time this book will appear, nearly six years will have elapsed since I discovered the voice of the œsophagus, and almost five since I published a preliminary account of this discovery in a book entitled *The Basic Law of Vocal Utterance*.[1] This discovery, though the most comprehensive and far-reaching of any that has ever been made, not only in regard to the voice, but in regard to the better comprehension of our nature and our entire human existence, has remained as unknown to the world as if it had never been made. Yet some day, when its importance is recognized, it will take rank in the annals of the history of the human race as second to no other discovery that has influenced and shaped human thought in the proper recognition of the origin and the nature of man, spiritual as well as physical, his abilities and his limits, and his relative position, influence, and destiny in the economy of the universe.

I have spent so many years of arduous labor on these investigations, and have become so thoroughly convinced of their truth, that I have ventured to make these assertions without the slightest compunction, or fear of final contradiction. Although the facts involved in these matters entitle me to these declarations, I would not have overstepped the bounds of modesty in so far as to make them had not my first experience forced upon me the conviction that the path of modesty in matters of this kind is not the one to success. I was so impressed with the exalted position of science, and so apprehensive of my own powers, that in my former publication I as much as apologized for my temerity in telling the scientific world things of which it did not have any previous knowledge.

---

[1] Edgar S. Werner. New York, 1894.

These last four years, however, have so enlarged my views and given me such a firm grasp and insight, that I no longer fear any man's judgment. I would, on the contrary, heartily welcome honest and competent criticism, being convinced that the same would not and could not but strengthen my position.

As a matter of personal gratification, I am indifferent to success; but I think the time has come when these matters should not continue to remain with me alone, but should become the property of all, not for my sake, nor simply for that of science, but for the sake of truth, and the benefit of mankind. Had my previous statements been given the consideration they deserved, other persons, in all probability, would have made *some* of the many discoveries, at least, that it has now been my privilege to make single-handed. Still, the field is inexhaustible; that which I have discovered being but an index hand to that which is still to be discovered. Having no reason to doubt but that I am a properly organized member of the human family, I consider myself entitled to speak of my personal experience as in like manner applicable to every other member of that family.

Having found it expedient to frequently address the reader in a "direct" manner, using the personal pronoun "you" in so doing, I must ask his pardon for this liberty. In thus addressing him, I trust we shall be in better rapport; all I shall have to say thus becoming, in a manner, a confession as from author to reader. While I confide in him and make him participate in these vital discoveries, I want him to confide in me, in so far as to take it for granted that all I shall say is truthfully meant, and that it has been arrived at, not superficially, but only after the most searching and long-continued investigations. We will thus become partners in a research as great as any that has ever agitated man's mind, or filled his soul with things of great moment. Having penetrated into matters which have heretofore been considered as occult, or inaccessible to man, my mode of proceeding will be found interesting as a guide to others wanting to pursue similar investigations.

In the beginning, it was all brought about by my simple desire, being a German, to speak the English language in the precise manner in which native-born persons speak it. For this purpose, I unwittingly pursued the same course which has been pursued by many others under similar circumstances; namely, that of introspection. Having been indefatigable in this course (which others must not have been), after pursuing the same for some time I was startled by unforeseen discoveries. They were phenomenal, and far beyond any previous design, hope, or expectation. After this, my original endeavor to speak the English language idiomatically correct became a matter of secondary importance. My eyes once opened, I *continued* to persevere in this course, and thus succeeded in penetrating deeper and deeper into matters heretofore deemed inaccessible to man.

Having pursued investigations by means of introspection now for a number of years, it has become an easy habit with me, and I can recognize and pursue processes by which results are obtained through *inner* motive powers, almost as plainly as such by which results are obtained through visible and tangible means. The facts thus observed and recognized as truths have become so numerous as to be almost overwhelming, in number no less than in importance; so much so, that I scarcely know where to turn or where to commence, to be able to communicate them all to others in due form and sequence. These facts are not temporary, but are constant; in so far as they can be conjured up at any time and under any circumstances, and are always of the *same* nature. They are of an entirely reasonable, practical, and, for the most part, mechanical nature; and are explanatory of the exercise of our faculties and functions, spiritually as well as materially. That these observations mirror actual proceedings going on within us for the production of vocal utterance, of breathing, motion, and locomotion, and the exercise of various other faculties and functions, it will be my endeavor, by actual demonstration, to prove through this and future publications.

For the purpose of enabling others to pursue a similar course of studies, I shall take especial pains to point out my course of proceeding as plainly as I can—such course with me having been entirely rational, positive, and direct, and without in any sense disturbing my ordinary mode of existence. The course pursued in physiologico-psychological studies, in fact, does not differ greatly from that pursued in the study of purely psychological subjects, which is also carried on by means of introspection, though it is of a more positive nature.

When the following was first written (it is nearly two years ago now), I intended, at an early date, to publish a short treatise on the subject of the voice only. Since then, however, the same has assumed greater and greater proportions, embracing many other subjects. Still I have deemed it best not to change this introduction in consequence thereof.

Though not quite ready for another publication (the subject is so great and my knowledge so inadequate), I do not know that I should have *ever* been *quite* ready, but for several incidents, all happening about the same time, which have induced me to break the silence I have observed since the publication of my book, *The Basic Law of Vocal Utterance*. These incidents, though in themselves apparently insignificant, have impressed me with the belief that I owe it to the public and myself to say something in explanation of what I have already said, and to add thereto (partly, at least) what has since been ascertained.

In the November, 1896, number of *Werner's Magazine*, I noticed the following:

"A good example of the inadequacy of expressional terms in discussing vocal topics is shown by Mme. Clara Brinkerhoff and Mr. Emil Sutro. Mme. Brinkerhoff has been a contributor to this magazine, and has addressed musical bodies, for many years. Mr. Sutro is author of the book, *The Basic Law of Vocal Utterance*. Both of them maintain that the voice is something more or other than an expiratory current of air set into vibration by purely physical agencies. Mme. Brinkerhoff thinks that the voice is the utterance of the soul, and that the soul has its seat in the solar plexus. Mr. Sutro scoffs at the theory that the voice is only out-coming air vibrated at or by the cords situated in the larynx. He thinks that the ligaments under the tongue also serve as vocal cords, and that speech is the product of vibrating ingoing air as well as vibrating out-coming air. Just what they think the voice is neither of these persons makes clear to others. Their failure to express their thoughts, however, should not be taken as proof that they have not caught glimpses of truths of the greatest importance. Still, our impression is that their concepts are too vague to be put into intelligible language even if the expressional terms at hand were adequate. But, all things considered, the fact still remains that discussion will continue to be largely useless so long as one person does not know what the other person is talking about."

In addition to all this, the proceedings of various societies in New York alone, judging by their reports also contained in the November, 1896, number of *Werner's Magazine*, which is of unusual interest throughout, show how great is the interest which, at the present time, centres around this matter of the voice. In place of saying the "truth" in matters of the voice, as contained in my book, it would, perhaps, be more correct to have said, "the first ray of light that has ever penetrated the gloom and the mystery surrounding the nature of the voice." In *Werner's Magazine* it is stated:

"If Mr. Emil Sutro's book, *The Basic Law of Vocal Utterance*, be right, then other writers on vocal science are wrong. His statements are startling and revolutionary. He claims to have discovered a new vocal cord and to be able to prove that speech sounds are the product of inspiration as well as expiration. The significance of this is apparent when it is realized that all vocal authorities, heretofore, have taught that voice is vocalized expiration, and that speech is this vocalized expiration articulated into words.

"The author draws a sharp distinction between the air taken for life-purposes and the air taken for speech-purposes. He says that vital breathing can and should go on independent of artistic breathing, and that the two processes need not and should not disturb nor conflict with one another. He combats the theory that the lungs are a reservoir of air, which in the vocal act is pressed against the vocal cords of the larynx, thereby producing tone, which is resonated and modified by the parts above the glottis. He maintains that it is a physical impossibility to give sufficient force and rapidity to the lung air to put muscular and cartilaginous tissue into tonal vibration,— that this force and this rapidity can come only from the internal atmospheric pressure, and that, therefore, preparatory lung inhalation for voice-purposes obstructs rather than aids the vocal act. He gives a new explanation of the formation of speech sounds, and offers various novel theories.

"Many readers will hesitate to accept his views, yet as long as vocal science is still in a formative condition and involved in so much chaos and uncertainty, any attempt at a solution should receive careful consideration."

I have cited this able review in full, written by one whose life has been one act of devotion to the solution of these questions, as it will at once introduce the reader into the drift of my investigations as far as they had advanced up to that time.

I have continued to steadily devote myself to the further prosecution of my investigations, never publishing anything, scarcely ever speaking on this subject to any one. The subject appeared to me so great and so far above my ability to master it that I, at first, looked around for assistance among those I deemed most likely to be able to render it. But no one had any assistance to offer, no one scarcely seemed even to comprehend what I was after. Thus, at last, almost in despair, I made up my mind that I must undertake this task single-handed; and I have been at it, scarcely without interruption, ever since.

Meanwhile, the play of "Much Ado about Nothing," or "The Farce about the Larynx," continued to go on bravely all over the world. I have watched it with a sense of pity, rather than amusement. It appeared to me, more than anything else, like a game of blind man's buff, in which *all* the participants were blindfolded; my own horizon, meanwhile, being illumined by roseate tints representing continuous new discoveries, like a May morn before the rising of the sun.

The voice has been treated as a separate mechanical issue, while it is the outcome of a series of both physical and spiritual issues. While the old school is reproducing, in its minutest details, the *dead* branch of a tree, I am portraying, in its majestic proportions, the broad expanse of a *living* oak.

These anatomical details may interest scientists; they are valueless to the singer, as he has no control over the movements of the larynx. He need but "attack" his note in the right way, and all these muscles, sinews, cartilaginous tissues, etc., will fall into line, involuntarily and unsolicited.

Now that I am offering innumerable *proofs* in corroboration of my assertions, I want scientists to take these matters *seriously*, and not to look upon this book, also, as some may possibly have felt inclined to do in regard to my previous publication, as a "scientific curiosity" merely. There are no greater problems before the world to-day than are treated here.

During all these years of unrequited labor, which extend far beyond the day on which I made my memorable discovery, my personal affairs meanwhile constantly suffering, with but one notable exception *no* hand was held out to me in succor. In view of this fact (and it is the experience of many who, in the privacy of their souls, are struggling after the light), I want to ask this question: With all the noble institutions for *learning*, why are there none to assist those who are attempting to solve questions *to be taught* for the benefit and advancement of mankind? True, there are scholarships and fellowships for students, but they are not available to persons advanced in years who have duties to perform and families to support. When successful in the end, their reward—if there is any—often comes too late to be of any practical value.

Such would be the case with me should any material acknowledgment come to me now, having of late attained to the leisure I had so much longed for, thanks to my previous labor and a brave son's devotion and valued aid and assistance. No man, however, will ever know how long I have been kept under the ban of purely materialistic endeavors, while these higher things were occupying my mind and clamoring for recognition. A sum equal to that representing a single day's expenditure for *falsely* teaching matters connected with the voice, alone, the world over, not to speak of other matters of still greater importance, would have sufficed for a number of years, if not for a lifetime, to place me in a position to devote myself exclusively to the exposition of the correct principles underlying these important subjects. As it has been with me, no doubt it is and always has been with many others in different fields of research.

Since the publication of my previous book, I have had four years of continuous experience, during which the statements therein made have been strengthened and enlarged, so that I am now ready to support them with an endless array of proof. That book, however, was the beginning of what some day will be regarded as a greater movement in the right direction than any previous one, for attaining an insight into nature's occult work in creating, developing, and sustaining the living organism, and the exercise of its faculties and functions, more especially *man's* faculties and functions. The subject, however, is of so subtle a nature that it cannot be treated like a mathematical problem or a chemical analysis; still, I shall do the best I can with such means as are at my command.

Recently an acquaintance who is interested in vocal culture asked me how I was getting along, and I answered, telling him something like what I have said in the preceding. He replied:

"That is the trouble with you Germans. This is a live world, a practical world; we want facts, results—something we can turn to account and make use of."

This impatience (and who can blame those who are suffering, or those who, being young and talented, want to be led into the right path) throws the door wide open to all kinds of charlatanism—charlatanism which is honest and charlatanism which is dishonest, the former, being more readily trusted, often working the greater harm. The best teaching for the present, in default of a science, is that which is based simply on experience; the pseudo-science now being taught being worse than no science at all.

While the exercise of speech is next to universal with all men, no one has any idea of *how* it is exercised; the wisest being as much in the dark as the least informed.

This is what so eminent a man as Oliver Wendell Holmes had to say on the subject in one of his lectures, delivered not many years before his death:

"Talking has been clearly explained and successfully imitated by artificial contrivances. We know that the moist membranous edges of a narrow crevice (the glottis) vibrate as the reed of a clarionet vibrates, and thus produce the human *bleat*. We narrow or widen, or check or stop the flow of this sound by the lips, the tongue, the teeth, and thus *articulate*, or break into joints, the even current of sound. The sound varies with the degree and kind of interruption, as the 'babble' of the brook with the shape and size of its impediments—pebbles, or rocks, or dams. To whisper, is to articulate without *bleating*, or vocalizing; to *coo*, as babies do, is to *bleat*, or vocalize, without articulating. Machines are easily made that bleat not unlike human beings. A bit of India-rubber tube tied around a piece of glass tube, is one of the simplest voice-uttering contrivances. To make a machine that articulates, is not so easy." [The Italics are Dr. Holmes's.]

It is not the *humorist* Holmes, however, who has said this, as one would suppose that it was, but it is the writer, scientist, and thinker, who was in dead earnest when he gave unto the world this "definition of the gift of speech."

Any comment on my part would but weaken the sense of the ludicrous this "explanation" of so great a subject, even from a mere mechanical standpoint, must arouse in the reader. Yet Dr. Holmes's "explanation" is not any more preposterous than that of many other scientists of the present day.

Teachers have said that, not being a teacher, I could not know anything about the voice. As if *they* had the sole patent right to the voice, and others held their voices but from them, in fee! I, however, took the liberty of looking into my own voice and trying to find out whence it came and what it was made of. It is not much of a voice, to be sure; yet it has the common attributes of all voices. Besides, I should like to know who, in truth, *is* a teacher. He who over a narrow path follows the footsteps of others, or he who strikes out boldly for the root and the truth of a matter, and, disregarding precedents, goes down to the very bowels of the earth, if need be, to bring it to the surface?

The knowledge of even the best of us is not much more than some froth on the surface of the well of truth. Yet that froth is all these timid souls have dared to examine. They have not had the courage to dive down deep into its fathomless flood. Many a truth has been taught by those who had been considered innocent of any knowledge thereof. I am one of these "innocents," and, on the whole, am not sorry for not having been imbued more with the knowledge, or supposed knowledge, of the present day.

We are so much the slaves of habit that we become reconciled to any condition, almost, no matter how undesirable or absurd it may be. Thus biological science has been going along in a rut for centuries, but little having been ascertained of vital importance; nor could this have been otherwise, considering the modes of investigation. I was not surrounded by so many trees that I could not see the woods. My perspective was as clear as a bird's, that soars above and beyond the smoke of the city and the dust in the eyes of the heirs of generation upon generation of anatomical and physiological research, burying beneath its lumber the clear insight of the soul. Thus, ignorance with me may indeed have been bliss. Yet I do not want to place myself in a position as deprecating science, having the highest appreciation for all its endeavors. I deprecate science only in so far as, dealing with matter, it attempts to draw inspiration therefrom as to spiritual issues; and the voice certainly is a spiritual issue.

The following appears in the *Encyclopædia Britannica*, under the heading of "Animal Magnetism":

"Mr. Heidenhain, after stating that in conformity with the manner in which one muscle is affected, others become similarly affected, proceeds to say: 'Probably the reflex excitement would extend still farther, but I naturally consider it out of the question to try whether the muscles of respiration would become affected. It is easily understood that such experiments require the greatest caution and may be very seldom carried out.'"

Valiant Mr. Heidenhain, brave explorer on a new and "dangerous" field of research. This is the *Ultima Thule* which any of these bold adventurers have endeavored to reach. *My work began where theirs came to an end.* Though I have not reached the "North Pole," I have gone far beyond anyone else.

COMMENTS OF A DISTANT REVIEWER

This entire subject is of so subtle a nature that I must warn the reader to be patient in its study and careful of his judgment. Should the present work, however, also fail to elicit the attention of my fellowmen, some thinker, perhaps, of a future generation, upon discovering a copy of this book on the dusty shelves of an antiquarian, while looking over its time-stained leaves and after struggling with its vernacular, may be struck with some remark coinciding with ideas arrived at by himself and other scientists of that day, and while commenting upon his "find," may possibly deliver himself thus:

"As the nineteenth century of the Christian era was drawing to a close, a citizen of the (then) youthful country of the United States of North America published a book which contained disclosures far in advance of his time and generation—truths, in fact, concerning life and the exercise of our faculties and functions, which, if properly understood, might have eventually led to even the solution of the very mystery of the soul. Though science at that remote period had made marvellous strides forward, its endeavors were mostly of a utilitarian character, or consisted of efforts to explain phenomena from a strictly materialistic standpoint. The author of this book, however, by dint of a combination of extraordinary circumstances, which induced him to search for causes of phenomena within, in place of outside of himself, had succeeded in breaking through the barriers which had, theretofore, separated phenomena which were called 'natural' from those which, by the majority of mankind, were still supposed to be 'supernatural,' or, at least, unexplainable, unknowable, beyond the ken of man.

"He was thus enabled to penetrate more deeply than any one ever had before into the knowledge of the mysterious forces which engender and sustain organic life. Had he been properly understood, the compass of human knowledge would have been greatly enhanced, and the race itself liberated from the narrow limits to which it had been confined by the scientists almost as much as by the theologians (by the doctors of the body almost as much as by those of the soul) of his day. Some writers of that period delighted in depicting a state of affairs several centuries ahead of their time. The changes which were supposed to have taken place, however, had reference to material developments only, and did not contemplate any advancement of a purely spiritual nature.

"Though the founder of the Christian religion, and other men of a high order of intellectual and moral insight, had laid down rules for 'deportment' which to a great extent still govern the world; in regard to a spiritual insight, the dearth, the waste, the discord, the distraction, the unrest, the 'Weltschmerz' (as the Germans called it), the despair of science, which knew but and dealt but with the baser part of our existence, unable to penetrate into the higher, was then at its height. The 'miracle' had ceased to exercise its influence over the intellectual classes, and knowledge had not taken its place.

"This writer, however, through his discoveries, had opened up the way—made a beginning—to a penetration of science into the realms of the spirit; and a substitution of faith based on *facts* for one based on tradition and fancy only. Religion and science, having been factors of a different, almost antagonistic, order, thus at that early period already might have become reconciled and united through *knowledge*; as to some extent, though by different means, they have become since.

"In thus gaining more knowledge, more light regarding the motive powers which govern our existence, the shackles which had overwhelmed the soul would have long since fallen to the ground, and a *truly* brotherly spirit would have prevailed among all classes and peoples in place of much of the prejudice, the insincerity, the overbearance, the animosity, the cruelty, and the insanity even of the believers in (or inheritors of) one spiritual theory (often misnamed religion) as against those of another.

"The world's thought, just previous to that time, had made great strides forward through the recognition of the laws of *evolution*, which culminated in one master mind, through great elaboration and by citing numerous examples, assigning cogent and necessary reasons therefor. The world should have been ripe, therefore, for this *greater movement* which it was now called upon to face; a movement which went beyond the mere recognition of phenomena and penetrated into *a priori* causes. Strange to say, it either could not or would not understand; being still bound by fetters which held it in a vise-like embrace of previously conceived ideas as to the impossibility of penetrating into matters of this nature, and which prevented it from even *testing* the numerous proofs offered by this writer as to the correctness of his assertions. His investigations, if properly understood, would have brought spirituality *home* to us; they would have made it accessible to us. It would have ceased to be a phantom, and would have become a reality, a friend on whom we could count, in place of a mysterious and incomprehensible stranger.

"Beginning with discovering the dual nature of the voice, the writer of this book opened up the way to the comprehension of the mystery of man's dual nature in *all* its relations. He made the discovery that the œsophagus is of equal importance with the trachea in carrying on the process of respiration and in exercising the faculty of vocal expression; that for these purposes œsophagus and trachea are to an equal degree directly amenable to the influence of the atmospheric air; that the dual nature of organic beings in general, and of man in particular, is represented by the hemispheres of the thorax and the abdomen; that the former in its entirety represents spiritual and the latter in its entirety material issues; that the trachea and its branches on the one hand, and the alimentary canal on the other, respectively represent these issues more directly; that the fusing and blending of these issues has for its result the phenomenon called life; that the severance of these issues has for its result the phenomenon called death; that there are thus positive limits, place, and surroundings assigned to material and immaterial issues within the sphere of our bodily existence, and that combined they pervade our entire system; that all phenomena of life, especially all phenomena of a spiritual nature, and among these more ostensibly those of vocal utterance, owe their origin to these issues momentarily joining hands; that in so doing there is a transitory fusion, which for an endless number of purposes is brought about in an endless number of ways.

"He discovered further that the larynx, previously supposed to be the *only* instrument for the production of sounds, has its counterpart in the 'replica' (the 'larynx' of the œsophagus), located beneath the tongue and represented by the frænum linguæ and surrounding cartilaginous tissues; that no vocal sound can be produced except by the coöperation of the larynx with the replica. He discovered the circulation of, and the origin of vocal sounds, and many other important issues.

"Through his discoveries, if properly recognized, *all* the sciences dealing with life would have been placed upon a new and far more reasonable and comprehensible basis than they had rested upon before.

"These discoveries would have tended to undermine the basis of every materialistic school of philosophy, and to place those with spiritual and ideal propensities upon higher and firmer ground. Had they been properly appreciated and further expanded by others it would have eventually become possible to develop *all* our faculties to the full extent of their ability, and to correct faults, errors, and defects caused by wrong education or heredity, through the application of laws at the very root of our existence; laws which were then, and in fact to a great extent are to this day unknown.

"It may, in fact, be said without exaggeration that his discoveries, which were all made within a period not exceeding five years, outweighed in importance all other discoveries combined relating to physiologico-psychical issues made previous to his time."

I can see many a reader smile after perusing the foregoing, and perhaps saying:

"Here is a Jules Verne of a new type come to deal with a novel subject."

Yet the time will come when the reader will cease to smile, and look upon these matters *seriously*. I do not mean, however, to throw down a gauntlet to science on these momentous questions in *a vaunting and reckless spirit*; but come as a petitioner rather, asking it to investigate.

My time and generation are but like a flash from the orb of eternity, but the laws I have discovered are as eternal as that orb itself. With all the scientific investigations now going on, there has not even an approach been made which might have led up to them; nay, not a hint or a hypothesis, even, leading toward the same. Science, in fact, had nothing to do with them; the first man might have made them almost as well as the latest. They are all grappling with matter, while I have grasped the spirit that is in, yet above, all living matter.

In making these discoveries I have bent a sail upon the crafts of physiology and psychology, which have been aimlessly, almost hopelessly, drifting on the shallow waters of the examination of isolated material phenomena. This sail will enable them to reach the broad expanse of the ocean, where they will be able to make soundings in its deepest waters.

Professor Huxley declared that during his fifty years of experience as a student and teacher not one thing really *new* had ever come under his observation. Had he lived to become acquainted with these facts I feel confident he would have declared them to be new.

The venerable Professor Virchow, the other day, in an address before the International Congress of Physicians at Moscow, made use, in substance, of these words: "The cell is immortal—there musthave been a previous cell for its generation. On this fact as a basis (ascertained by the aid of the microscope) the science of the coming century may securely rest."

And he set this down as the greatest achievement of science in respect to the recognition of the phenomena of life. Yet there is nothing more fallible than the microscope in ascertaining facts regarding the knowledge of life. It may to some extent reveal the essence of *matter*, but it is not given to it to assist in recognizing the principles which govern life and the *spirit* of life.

FRAGMENTS

This book, in a sense, is a personal narrative, and necessarily must be so, giving an account, as it does, of observations in experiments upon myself. In making these experiments I have endeavored to treat myself impersonally, as a subject, so to say, placed at my disposal for experimental purposes; my ego having been the object as well as the subject of my investigations. In occasionally speaking of the results thus obtained in a eulogistic manner, this should not be looked upon as self-praise, therefore, but rather as an impersonal mode of describing what has come under some one's observation—this "some one" being myself. I want to place the matters I have observed before the reader in the right light, and do not hesitate to say or fear to say just what I think to be the truth. If I were to wait for others to say these things the reader who does not comprehend their latitude as I do might have to wait a long time before he could grasp the subject in its entire importance. I want to say this much as an apology and a vindication for frequent indulgences in apparent self-eulogism.

I have another motive for making such remarks; viz., the desire of rousing the scientific world from its apathy regarding these matters. These laudatory remarks may wound its pride, and possibly arouse its ire,—more especially in view of their coming from a layman,—and thus induce it to study these matters, if but for the purpose and with the view of controverting them. I would hail such an endeavor with pleasure, not having the slightest fear of its ability to successfully controvert any of the vital facts I have ascertained, and whose correctness I expect to prove by a great array of facts with accompanying proofs.

When I first began to make these studies, I made numerous notes as new features happened to present themselves to my mind. I have encountered no inconsiderable difficulty in sifting this material so as to present my experiences in as connected and consecutive a manner as possible. In this, however, I have only partially succeeded; nor have I been able to altogether avoid repetitions. For these shortcomings I must plead a want of time. For some time past, however, my experiences have accumulated so rapidly that I have ceased to take any notes whatever, trusting to my memory that these mental notes may be recalled at the proper time. No doubt some things, even of importance, have thus been lost sight of. Still, while pursuing similar studies, they may in the course of time turn up in some one else's mind.

In looking over some of my notes I have found things which I have deemed worthy of preservation. I let some of these follow in a promiscuous manner. This, it must be admitted, is not in accordance with scientific usage. But I am not a scientist, simply an amateur; and take advantage of the privileges this fact gives me. If I were to conform to strict scientific rules and "etiquette," years might elapse before I could get these matters into proper shape. It will always remain a mystery to me, however, why these things should have come to me at all—so unworthy, so unadapted to their proper exposition. In order to do them justice, they should have come to one complete master of his time, young, strong, possessed of a wide range of knowledge and a deep insight.

I will now let follow some of the matters I have spoken of:

My personality and my work must go together, until others relieve me of the latter by making it *their* work to the same extent that I have made it mine. You cannot separate the fiddle from the fiddler, neither having any significance apart from each other, except by the fiddler perpetuating that which the fiddle produces—the composition,—by writing it down, thus transmitting it to others. This I am trying to do by this book.

No doubt some of the things which have come under my observation in some form or other are already known to science, and are, therefore, a corroboration, or an explanation, only, of things already known. With me, nevertheless, *all* is original; and I may therefore justly claim that if any of these matters have been discovered before, I, at least, have *rediscovered* them.

If I were an institution possessing a guaranty of continued existence I might value the present lightly, knowing a future would come when these matters will be fully understood. Being a creature of the present, however, which may be turned into the past—especially at my time of life—at almost any moment, these matters should become known at the earliest opportunity; some of them being of so subtle a nature that they may require personal explanation and illustration. They have been hidden from us in the past; should they fail to be made known now, *the same opportunity may not arise again for centuries.*

***

I do not claim any special sagacity over others for having made these discoveries, and disbelieve altogether in miraculous interposition. Yet I do not want to be prejudiced in any direction.

We are surrounded by the mysterious and the miraculous; and that which is called "natural" as a rule is far more mysterious than that which is called "miraculous."

"Truth is stranger than fiction"; which is undoubtedly true. We can imagine that only of which we have at least *some* knowledge, but there are realms of truth beyond us of which we have *no* knowledge. Besides, these revelations are of so extraordinary a nature that I cannot altogether close my eyes to the fact that I *may have been led on to them* by agencies beyond my personal power of volition. I will cite but one reason why such an idea might be justly entertained by me.

That which originally led me on to these investigations, as already mentioned, was the simple desire to speak the English language just as native-born persons speak it. Although I eventually became aware of the fact that this was next to impossible, yet I persisted in this endeavor to such an extent that I spent far more time on it than it would have deserved had I been *convinced* that I would be finally successful. Again and again I said to myself, "This is a foolish, absurd, unworthy undertaking for a person of intelligence"; the next minute I was at it again, trying to utter this sound or pronounce that word in the "correct English fashion."

I want to ask, What was it that impelled me to thus persist, almost against my wish, will, and better insight? When, after many years of this almost wanton endeavor, I discovered the dual nature of the voice, I could not help but think that an influence beyond myself had been exercised to impel me to persist in these efforts, which were then crowned with a success of a different order, and far beyond any previous expectation. *I then found what I had been after unknown to myself.* To simply say I was "infatuated" would not explain this strange adherence to what for a long while looked like a vain and hopeless undertaking.

I am aware that for me to say, as I have just now said, "I cannot altogether close my eyes to the fact that I may have been led on by agencies beyond my personal power of volition," may expose me to ridicule in the eyes of some persons; besides being a contradiction to my other convictions. Yet I say so deliberately and am quite willing to abide by the consequences. It is a case of the duality of our nature, which impels me to take a naturalistic or biogenetic view of things in one direction, yet forces me to take a spiritualistic or abiogenetic view of them in another direction. I do not comprehend those who under *all circumstances* are capable of pursuing either the one direction or the other.

***

I might say I have been on a prospecting tour to a *new* country, where I found the outcroppings of numerous veins of precious ore. These veins are *true fissure veins*, penetrating, as they do, into the very bowels of the earth; and it will take centuries to exhaust them in all their *dips, spurs, and angles*.

***

It will be a matter of surprise that a layman, one not of the tribe which make science the pursuit of their lives, should have penetrated into these mysteries. It must not be lost sight of, however, that science, as a rule, deals with things visible and tangible, while the voice is a sensation which, regarding its origin in the ego, cannot be observed outside of the ego. One may by close observation trace the origin of one's voice to its innermost channels, and thus learn much about the subtlest characteristics of its nature, a proceeding to which it would not be possible to subject any one else's voice. The same conditions prevail in regard to other sensations which have also come under my, at least, partial observation.

***

Science, as a rule, has been satisfied with the observation of results, of phenomena, without attempting to penetrate into causes, which seemed to be unalterably hidden from its gaze. Special features, however, of the voice have been ably and successfully observed and described by many eminent persons. To these I have not given any attention, partly because they were beyond my sphere, and partly (not being a musician) because they were beyond my power of observation.

***

In looking for the voice, anatomy in its minute examinations of the larynx has but opened up a grave for us to gaze into. And what have we beheld? A skeleton of the voice's body—of its soul not a trace. This skeleton, to boot, is but a *portion* of the mechanism of the voice; of its other parts, equally important, science has not even known that they were in existence. Like a palæontologist or an archæologist, I have dug up these other parts or fragments from all around; some were found close at hand, others quite a distance off. I have skilfully put them together, and have thus constructed a fairly *complete* torso, or framework of the voice. I say "torso," though I may justly claim more than that, having again infused the soul into it which had fled from it; and, see, it has become a *living thing.*

That the wonderful apparatus contained in the throat is for a purpose there cannot, of course, be any doubt. It is but partly for the purpose attributed to it, however, and, until we better comprehend this part-purpose, especially in view of the fact *that we have no control over its mechanism,* it will be best, as far as singers and elocutionists are concerned, to surrender it to and leave it with the anatomists.

To the ultimate aim of science—the knowledge of life—I have contributed matters of a nature deemed beyond the province of the knowledge of man. Was it ever intended that they should be known? On more than one occasion I have been puzzled to know whether to go on with these investigations; whether I had a *right* to go on with them. Still, I was sustained by the fact that I had been *led on to them.* For what other purpose could this have been done but for that of making the results thereof known? They could serve no good purpose in remaining locked up *within myself.*

It is my belief that the ordinary course of events is never interfered with; but that *great* events may be inaugurated by unseen agencies and guided by unseen hands. The responsibility which has devolved upon me, incompetent and unprepared as I am, is almost too great; still, I must try to discharge it to the best of my ability.

***

I have no personal motive of either fame or fortune. At one time I would have been pleased with such results; now it is too late. If not in my day, some day, I trust, some one will read and comprehend; some one will not mind the trouble of investigation. It is not likely that I shall *forever* remain the only "seeing one."

It would have been better if I had not published a line for at least ten years. It would have taken that long to say what I want to say, *properly*. My time is too uncertain, however, to run such a risk. My friends are falling to the right and left by the roadside. I must be up and doing; must make a beginning at least.

We must be satisfied with reaching matters approximately, and argue by analogy to some extent; and also hope that others will take them up and push them along a little farther than we have been able to do. Perhaps in the course of time a perfect insight may be arrived at.

***

The community of man is a necessity; a separate existence, an anomaly. We are dependent and interdependent upon one another. Man cannot escape his fellow-man. In the remotest desert his spirit is still in communication with him. If it were not so, who would not at times want to flee all, escape from all?

I have but one fear—inability, for some reason or other, to finish my work. I feel like the heroine of a celebrated German novelist, travelling about with a trunk filled with gold, which she distributed among the *deserving poor* as fast as she came across them. Meanwhile she was in constant fear lest her life should ebb out before all was distributed, and its precious contents *lost* to those for whom they were intended. If there were any way of imparting this knowledge other than by writing it down, I would gladly resort to it. But how can I reach the few who are capable of and willing to take up these questions, except by communicating them to the many? These "few" will be found in all parts of the world, for these truths apply to *all* men, independent of sex, race, or country.

***

My cry is not for recognition. My personality might be blotted out, like that of millions of others, without its being noticed, yet, by virtue of this trust which has been reposed in me, what a loss it would be! My cry is for investigation and the coöperation of others, so that this work may be carried on independent of myself. Meantime, I cannot transfer this task to others. I must first explain all that it is in my power to explain. I can then shift it from my shoulders onto theirs. They must be educated up to it before they can take hold of it as I have taken hold of it.

***

When I first announced my discoveries, I gave all I possessed, supposing others would see as I saw and comprehend as I did; having no doubt but that the world would at once acknowledge their truths and accept their precepts. I have since found that the world can get along very comfortably with a vast amount of want of knowledge. I therefore made up my mind not to be quite so rash again in making it my beneficiary, not till I was better prepared for the purpose; this other book of mine having been finished rather hastily in the erroneous belief that this knowledge was at once and imperatively needed.

Since publishing this previous book I have also found, which I did not know at that time, that my very mode of investigation (by means of introspection) was new; that no one had ever looked into matters of this kind in the manner I had; besides, it seems strange that in this age of keen investigation of the most trivial matters, no one should have deemed it worth his while to look into these more important subjects.

Regarding the anatomical investigations of the larynx, and anatomical, coupled with physiological, investigations generally, let me ask a question: Supposing a palace with a million apartments, each one in succession more luxuriously furnished than its predecessor, would they avail anything to its *sole* inhabitant, if that inhabitant were blind?

We have obtained a fair conception of the wonderful palace, the human body, its numberless apartments and their luxurious furnishings, but do not comprehend their meaning, except in a remote and unsatisfactory mechanical sense. *We* are the blind that inhabit it. Most of these apartments will remain meaningless to our understanding until we ascertain what use the sovereign, the soul, which reigns therein, is making of them, not only mechanically, but *spiritually* as well. For the soul lives in them all, though it is supposed that it lives only in its throne-room of the brain and that it never descends from the throne set up in the same.

Just here biologists have blundered, trying to get hold of *psyche* by pursuing matter bereft of life; or investigating life in other beings instead of that inherent in themselves. The vivisection of all the frogs in the world will not give us the first knowledge of the frog's soul; certainly not of *our* soul. The knowledge of the anatomical construction of the larynx has brought us no nearer the knowledge of the mystery of the voice than that of the brain has brought us to that of the soul. We must understand the process by which the mechanism of the brain is set in *motion* before we can begin to understand our mode of thinking. We must comprehend the manner in which a musical instrument is to be used before we can begin to draw music from the same. And so must we understand the spirit which moves the mechanism of the voice (of which so far we have known but a single factor), if we want to understand our mode of using it.

Does any one seriously think that by photographing vocal sounds, or passing a mirror down his throat and watching the movements of the vocal cords, he will observe anything that will lead him to an intimate knowledge of nature's subtle process by which vocal sounds are produced? As well look at the face of a clock and see its hands move, and then say you have arrived at a knowledge of the hidden intricate mechanism of the works of the clock. The mechanism of the instrument of the voice is a thousand times more intricate than that of a clock. It lives, it breathes, it moves, it expands and contracts, it rises and falls, it gathers, it gives—now here, now there.

Starting from the supposition that life is too subtle, too intangible a thing to have its innermost operations disclosed by the clumsy work of our hands or the dull vision of our eyes, though increased in power a thousandfold, I matched the subtle work of my voice with the subtler of my brain, and thus, undisturbed by any extraneous agency whatever, watched the process by which, first, simple mechanical, then articulated sounds, and finally sounds linked together into speech, are produced. In so doing I traced sounds through the labyrinth of numerous avenues to their original sources—*the organism of all our faculties, instead of being confined to their end organs, being widespread over our entire system.*

***

Physiologists as a rule are satisfied with the *observation and exposition* of phenomena. I have endeavored to *explain* phenomena. I have gone "behind the returns," as politicians say. I have lifted the mysterious veil, and have obtained glimpses at the process of life. In this manner the voice of the œsophagus was first discovered, which, in logical sequence, has carried me from one discovery to another. Once in the confidence of nature, it freely opened up to me its heart. Comprehending one thing led me on to the comprehension of others.

There is no study which is as fascinating as that pursued by introspection. It is self-compensating in the highest degree; all facts thereby evolved being the logical sequence of others previously ascertained. Or, if not always in sequence, they all fit into the same system; everything that has been ascertained being a stone which was waiting to be placed in a certain niche to fulfil a certain purpose in the construction of a harmonious edifice. There was no waste, no material entirely lost; nor will there be at any future time. If similar studies will be pursued by those specially fitted for the purpose, the time may not be far distant when there will not be an atom of our material existence whose meaning and purpose will not be understood. The laws which I claim to have discovered will assist in this accomplishment, as they are of so broad a nature that they may be said to form the substructure to forces and conditions which are at the very root of our existence. I do not pretend to say that in this little book they have been properly treated, nor that I possess the ability, under the best of circumstances, to thus treat them. I have but stated what has come under my observation, and have stated it in as simple and direct a manner as my instinct and my ability have taught me to state it.

I have been up on Mount Washington to see the sun rise. It was a beautiful picture; still, there were clouds in the way which here and there obscured my vision, as was to be expected from the unwonted height to which I had risen, and the distant horizon.

***

I am not writing for a class, but for the multitude to which I belong, and of which, in its aspirations, its hopes, its sincerity, and its ignorance regarding *specific* knowledge, I form a part. Hence mythoughts are its thoughts and my language its language. There will be no difficulty, therefore, for *all* to understand me and to profit by my experience.

My observations result in the triumph of the sensation, the feeling (common to all), over the exact sciences (known to but few). Science, for the most part, is satisfied with dissecting or analyzing. My endeavor has been to construct; to form the whole out of parts instead of reducing the whole into parts. My guide has been instinct coupled with common-sense,—that rarest of all the senses in spite of its name. How far it has guided me aright, it will be the province of science to judge.

I may be asked why, in treating upon so "simple" a subject as the human voice (my only endeavor in the beginning), I want to move heaven and earth, and press them into my service. My answer is, Wherever I touched the subject of the voice, I found it to be in correlation with all other subjects.

My great desire now is, that I may be granted the time and retain the ability to write out all I have ascertained; while my greatest wonder is, that these things should have waited for me at all to be made known; why they should not have been discovered centuries ago. My eyes once opened, I found them lying about within the easy reach of my arm and the mere assistance of my pick and shovel, like precious ore in a newly discovered mining country. I had but to open the lid of the mysterious casket which had been intrusted to me, and all these great truths escaped from the same; not to disappear, however, as they did in the fable, but to remain with me and to be made known through me to the world.

***

The best part of my life has been spent in this, my adopted country. Though I experience no difficulty in expressing myself in the English language, still it is not my native tongue, and I sometimes feel as if I might have said some things better if I had said them in German.

**

Looking at the many volumes written on the subject of the larynx alone, and considering that during all this time its associate, the replica, without whose assistance *not one* vocal sound can ever be uttered, has remained unknown, though in plain sight and "in everybody's mouth," one cannot help but think of Goethe's lines:

"Ein Kerl der speculirtIst wie ein Thier, auf duerrer HaideVon einem boesen Geist im Kreis herum gefuehrt,Und ringsumher liegt schoene gruene Waide."

("A theorist is like unto a beastOn barren soil by evil sprite led round and roundWithin a narrow circle, though beyond there is a feastOf pasture green on fertile ground.")

"THE BASIC LAW OF VOCAL UTTERANCE"

My earlier work, entitled as above, was written under peculiar circumstances. After discovering the fact that sounds proceed from beneath as well as from above the tongue, light streamed in upon me on so many subjects I had previously attempted to solve that I was almost dazed thereby. I thought it my duty to make these matters known, and attempted to describe them as they appeared to me. They were all perfectly clear to me, and even to-day there is scarcely a thing I then said that does not wholly stand its ground. Still, to-day, viewing things from an advanced point of view, much of that which was then expressed pragmatically, almost in a single sentence, and which then appeared to be sufficient, I am convinced requires considerable elaboration and elucidation.

Take, for instance, this dictum: "The manner in which we breathe for speech is by raising and lowering the tongue," etc. This is perfectly correct, and positive proof will be advanced hereafter as to its being so.

I thought these matters would be readily understood, not knowing at that time that between the manner in which I had reached conclusions and the one in which conclusions had been reached by others who had also made a study of these matters, there was a vast difference. Unknown to myself I had lived a life of my own. I had given myself up to these matters in a manner no one ever had before; having been everlastingly at it, holding on with a tenacity that knew no restraint. In this manner I wrung facts from nature that may have never been intended to be revealed.

There was something Faust-like in it all, and I sometimes shudder at my own temerity. Still, I had no such thought when

I so persistently continued trying to fathom the mystery of vocal sounds. Viewing it in its proper light it was a narrow and every-day undertaking. I was fairly staggered, therefore, when I reached such unlooked-for results.

The reader, however, may ask, and I feel it incumbent upon me, as well, to tell him, What was the nature of these results? Wherein consisted these discoveries? They covered a large field and whole range of knowledge. They had reference more particularly to vocal sounds. These, in fact, had almost exclusively occupied my mind for many years. These apparently simple factors, vocal sounds, I have since ascertained are the outcome of laws, forces, and agencies, and combinations of all these, which largely make up the sum and substance of our spiritual existence. The direct nature of vocal sounds, therefore, cannot be well treated upon till some understanding has been arrived at of the nature of the elements out of which they are composed. I was rash enough to attempt to explain them, especially the consonant sounds, in this little book of mine, from a standpoint I had then arrived at. Others have tried to explain them from a much narrower standpoint still. From that standpoint I offered explanations as to our mode of speaking, breathing, as to defective speech, etc. Although this was an advanced standpoint, and well worthy the consideration of scientists, it was a standpoint far beneath the one I have arrived at since.

In attempting to scale a mountain I had reached a point from which I could overlook the valley immediately beneath my feet. I have since gone up much higher. Yet there are towering heights still above me which I shall never be able to reach. From this it will be seen how difficult it would be for me to state in a few paragraphs what I had actually ascertained. That book, however, will increase in value in the course of time, not only for the knowledge it contains, but historically, so to say, as the beginning of an evolution which, it seems to me, will eventually embrace all sciences in regard to man; when treated, as they will be, from a standpoint of inner as against one of outer consciousness, from the standpoint of the soul and the heart, as in the inadequacy of our expressions I have to call them, as against that of the head and the senses.

I have since arrived at a plan according to which these matters will be treated in a more systematic manner. In *this* volume, besides many novel subjects, I have been enlarging upon and elucidating many superficially mentioned in my book, *The Basic Law of Vocal Utterance*. Still, the matters treated upon even in *this* book cover so much ground, and had to be condensed to such an extent, that many of these also will require further enlargement and elucidation. This will be attempted to be done in future publications. Meantime I trust these matters will be taken in hand by others, who by their writings will relieve me of some of this additional labor. Take it all in all, there is so much of this work that I feel as if I had swallowed the ocean and was now called upon to give an account of its contents.

THE VOICE OF THE ŒSOPHAGUS AND ITS VOCAL CORDS

Among the discoveries mentioned in my former publication one stands out most prominent, and it is the basis of all my other discoveries; namely, "that the voice is of a dual nature." I had ascertained that sounds circulate around the radix of the tongue; that they, or rather the air wave which carries them, enters either at the upper surface of the tip of the tongue and recedes back, to come out again from beneath its lower surface, or vice versa. I had also ascertained that the former process is the English, the latter the German, for breathing and vocal expression.

I was convinced that this signified a circulation of vocal sounds; and though I had finally also reached this conclusion and intimated it, namely, "that we breathe and speak through the œsophagus," I did not express it in so many words, as I meant to leave this expression for a future publication. I was at first under the impression that both waves belonged to the trachea, the one that was ingoing as well as the one which was outgoing.

Meantime I had discovered the "larynx or voice-box to the œsophagus," but considered this at first also as belonging to the trachea. I thought inspiration and ingoing sounds belonged to the vocal cords of the trachea, expiration and outgoing sounds to this "new" vocal cord located beneath the tongue. To study these first attempts, by which I was trying to find my way, and which culminated in these wonderful discoveries, I presume would be of interest to the student. I can here mention only the main points.

I have found beyond a doubt, and my future statements will more fully establish this fact, that the frænum linguæ and the parts of the mucous membrane surrounding the same are relatively of the same nature in regard to the voice of the œsophagus that the vocal cords and other parts of the larynx are in relation to that of the trachea.

In contradistinction to the larynx, I named these entire surroundings the "replica," as, in conjunction with the tip of the tongue resting upon the same, they conform to the shape of the oral cavity, of which in their general appearance they are almost a counterpart. In a similar manner I named the special part thereof, which "regulates" the intonation, the "vocal lip," in contradistinction to the vocal cords of the larynx, which perform the same service for the voice of the trachea.

After making such positive assertions regarding the replica as I did in my previous publication—now more than four years ago—I was more than surprised that no one should have deemed it worth his while to look into the value of these assertions. If any one had, he could not have helped but acknowledge their correctness. It is but necessary to utter any vocal sound whatsoever, either vowel or consonant, and while doing so watch the vocal lip and the frænum, to become at once convinced that their motions are of precisely the same order as those of the larynx and the vocal cords.

So many have spent year after year upon the difficult and "fruitless" endeavor to study the motions of the larynx; while here is an opportunity plainly before every one's eyes to study, without effort, the most interesting phenomena in voice production. We must be obliged to seek for a thing high and low before we deem it worthy of our attention.

THE HUMAN VOICE

What is the voice—a spirit, or "an expiratory current of air set into vibration by purely physical agencies"? It does not seem to me to be either, but something which is of the nature of both: our dual nature, embodied in the sounds of speech; our body and soul joining hands to produce the miracle of the voice. Regarding the materialistic view quoted above, which is held by most of the investigators, who make the larynx their *point d'appui*, I think that if there is anything in our composition or emanating therefrom that is *not* produced by "*purely* physical agencies," it is the voice.

In my opinion there is nothing purer, more "spiritual," in the world than a beautiful voice. Did you ever *see* a spirit? Perhaps not. But you have often *heard* one. You hear them daily, hourly, constantly; other spirits as well as your own—the spirits represented by the voice; the soul incorporated in the sounds of speech. When you converse, it is soul to soul; when you hear an anthem sung, it is the soul of the singer to the soul of the universe. The soul reveals itself most prominently through the voice when there is anguish in it, or joy; tears or laughter; love or hate.

An attempt to get at the truth in matters of the voice is an attempt at getting at the truth in matters of life. If you will tell me *all* that a vocal sound is, I will tell you what your soul is.

To examine into the anatomical construction of the larynx, to watch it physiologically and learn to understand the motions of the vocal cords in their relation to vocal sounds, is not much more than looking at the dial of a clock (a simile already used, but worth repeating). The movements of the hands will give you *no* cue to the construction of the intricate works hidden behind the face of the clock. Nor will the careful examination and observation of the "dials" which serve the voice of the œsophagus in the same manner as those of the larynx serve the voice of the trachea, measurably increase the knowledge of vocal phenomena. I do believe, however, that, inasmuch as the movements of the replica, the frænum, and the vocal lip fit into and complement those of the larynx and its vocal cords, and vice versa, lessons of great benefit to the knowledge and the improvement of vocal utterance may be learned, *after* we have once begun to understand what these movements imply.

That we cannot now derive any benefit from the observation of these motions is due to the fact that they are *reflex, involuntary, uncontrolled* and *uncontrollable* by the will. Or, as Mme. D'Arona expresses it:

"They are not the *cause* of the perfect tone, but are simply acted upon by the cause."

After having become acquainted with the cause of these motions, and having learned to control it in the interest of pure and perfect tone, the movements of the larynx and the replica will become of value to us as "indicators" of the correct or incorrect exercise of the cause which they reflect. In "recording" the original movements they will show us what is right or wrong in the latter, and will thus offer us an opportunity for correcting them. Up to the present they have been simply barometers, which, no matter how closely we may observe them, offer us no opportunity for changing "the state of the weather" which they indicate. After thoroughly comprehending the *causes*, however, which move them, we may shape the course of the latter in conformity with our will. Or, vice versa, we may shape our will, which, after all, is the *first cause*, so as to correct that which they indicate to be wrong in our tone production.

Now, what is that which the will acts upon, and thus becomes the original source, the first cause, so to say, of tone production? My answer will be a surprise, for, as far as I know, no one has ever as much as thought, even, of looking in this direction for the seat of the voice.

The original source of tone production has its location in *various vessels of the viscera*: in the lungs, the kidneys, and the bladder, for the most part, though many other vessels, if not all, participate, and are more or less involved in its production. Besides these vessels, the heart and the solar plexus, as central organs of the vascular and nervous systems, together with the brain as the central seat of thought and the will, perform parts of the highest importance in tone production and vocal utterance. In the lungs, the bladder, and the kidneys, together with their coadjutors, the bronchi and ureters, *the tone originates*. Here we can control, and unconsciously do control, it.

I shall adduce indubitable proof as to the correctness of these assertions. More than that, I shall *locate* sounds in these various vessels. As a tone proceeds from a given string located in a given part of a musical instrument, and cannot proceed from or be produced on any other string, a given tone of the human voice proceeds from a given vessel, and cannot proceed from or be produced in any other vessel.

I shall furthermore show that the various shades of a tone proceed from various parts of such vessel. Yet, while tones are produced in special parts, the instrument of the voice being of a sympathetic nature, *all* parts of the *viscera* participate therein, by, in a manner, *leaning* towards a vessel in which a tone is produced, thus assisting in giving it utterance. If a sound is produced in one of the vessels of the abdomen, those of the thorax, though not directly participating therein, give it aid and comfort by their passivity, thus throwing the entire strength of the voice-producing forces into this one spot. If a sound is produced in the thorax, the vessels of the abdomen aid it in a similar manner. This is more particularly the case when a sound of a superior order is to be produced, which is thus *reinforced* by this aid.

In matters of the voice, as in many others, truth is stranger than fiction.

Dr. Rush has said:

"Some day, when the real instrument of the voice will be discovered, it will be found to be of an order far different in its nature and construction from that which it has ever been supposed to be."

The greatest mechanical wonder, however, is that the voice, and that which is apparently one and the same sound, should under different circumstances emanate from sources so entirely different in their construction as the vocal cords to the trachea and those to the œsophagus, the viscera of the kidneys, the bladder and the lungs, etc. This fact also accounts for the mystery which, like an impenetrable veil, has hung over the features of the voice. Who has ever thought of looking for the spirit of the voice to reveal itself from *beneath* the tongue? Who has ever thought that the œsophagus was a breathing-tube of a similar functional order as the trachea? Who has thought that the viscera of the abdomen were playing as important a part in breathing as the lungs? Who has thought that the hemisphere of the abdomen was as directly amenable to the influence of the air as that of the thorax? Who has, in fine, thought that the viscera of the abdomen together with those of the thorax were primarily instrumental in producing the voice and vocal utterance?

It may not be pleasant to know, and it may not quite conform with our æsthetic taste, that the "voice divine" should have its origin in such vessels as the kidneys and the bladder; but I have no quarrel with the Creator, and can but wonder, as I have never ceased to wonder from step to step in all these investigations, at the marvellous resources of nature. There is one great lesson conveyed through this, namely,—- that the body is *divine* in its *every aspect*; parts which have been supposed to serve ends only of a comparatively low order participating in the highest spiritual functions.

This knowledge is the sanctification of the "flesh," so constantly and unjustifiably rejected and reviled as against that of the spirit. I am not dealing with theories, but am stating facts which will be as positively proven as any other scientific facts ever have been proven. These proofs will not be all forthcoming in this book, however, there being other subjects of equal, if not greater, importance that I have to deal with before I can reach them; these subjects being of such a nature that they must be explained before those immediately connected with voice production can be properly dealt with.

I have been reproached with attempting too much; with dealing with too many subjects at one and the same time; that I ought to complete one theme and then take hold of another. Just so; but this cannot be done. I must first deal with general principles. Our entire system being of a homogeneous nature, I cannot deal with separate issues until these principles have been dealt with and understood in their entirety. Besides, I cannot hope to ever *complete* any one thing. I shall be well satisfied if I shall be able to simply touch upon every subject that has come under my observation, lightly, suggesting things, and leaving it to others to enter more thoroughly into the same.

INTROSPECTION

With our mortal eyes turned outwardly we cannot see spiritual things, nor the motive power of life, nor the material form the spirit assumes in moving the mechanism of the body. For there *is* a material way in which it is thus moved, as there necessarily must be, and I have obtained glimpses thereat by turning my eyes inwardly—by looking into myself with the *inner* surface of my eyes.

Yet through all these centuries people have been using that portion of their eyes which is intended for external vision only, in a vain endeavor to arrive at spiritual-material facts. Thus the larynx, as the supposed seat of the voice, has been subjected to scrutiny based upon laws derived from phenomena which owe their origin to physical causes only. During this vain endeavor the larynx has been subjected to torture and maltreatment worse than that inflicted upon a mediæval witch.

But its tormentors have derived no solace from this treatment, not even that of a confession of imaginary sins. Why not? Simply because it had not anything to confess, being a reflex, an indirect, and not a free and original agent. Through torture (by means of the laryngoscope), the destroyer of harmony, we cannot arrive at laws based upon harmony.

Is not all physiological research more or less of this order? The "higher law" of science may demand its victims, even as did the "higher law" of the church. I do not wish to say, however, that the sacrifice of animals on the altar of science is as useless as that of human beings used to be on that of religion. Vivisection, however, while it may, and no doubt sometimes does, help to recognize the physical cause of disorder, will never be of any value in arriving at spiritual causes and the recognition of the inner motive power of life, nor to any great extent at that of the exercise of our faculties and functions. For this knowledge we require a different mode of proceeding. To penetrate into the realm of the spiritual-material world (and all phenomena of life are of that nature) we must not look externally but internally, not into other beings but into ourselves. That is the only place where we can hope to find it in action and arrive at the causes of such action.

As our being cannot enter into the inner life of another being and identify itself with the same or become a part thereof, or remain apart and become a spectator of the same or substitute therefor (not even for that of the simplest and lowest living vegetable or animal organism), we would have to despair of our ability of ever being able to arrive at the laws governing life, if we were not able to look into our own lives by substituting for our observations our inner for our outer consciousness.

The word "Introspection" has heretofore meant reflection upon purely spiritual phenomena only; I have proven by my personal example that we can observe physiologico-psychological phenomena with considerable accuracy—very little of this kind of work, as far as I can learn, ever having been done before. The nearest approach at amalgamation, probably, is that which is brought about by means of hypnotism. In this instance the two factors, the positive and the negative, the operator and the person operated upon, do not fuse, however, and become one, but remain entities, each in his own right. Or, to speak still more to the point, while the positive, that is the spiritual, factor of the operator may, and no doubt does, join hands with the negative, that is the material, of his subject, by which the operator becomes one with the latter, there is still but an *influence*, and not an insight. Besides, this condition is as yet too obscurely known to be made use of as a practical means of observation.

After all this, the question will still be asked, "What must we *do* to look into ourselves?"

I will admit that I have not stated what others should do, but in explaining what I have done I mean to explain what general course others will have to pursue. By taking into consideration what I have said, and adding thereto what I shall still have to say, a general idea may be formed of what the reader must do to place himself in a position to make original observations by means of introspection. No two cases being just alike, from the fact that heredity, the mental capacity, physical condition, education, temperament, nationality, etc., with no two persons are just alike, it is not well possible to point out a course quite suitable to all. I might as well attempt to arrive at a law by the observance of which *all* persons would be enabled to write poetry.

Still, needing assistance in this vast undertaking, I am particularly anxious to make this matter clear, as the results of these observations are of vital interest to all, and I am but one weak, ignorant mortal creature, with but a small fraction of a life left to me in which to state that which it would at least take a full lifetime to properly and fully explain. I am overburdened with an insight which is being increased daily, even against my will, and which I shall never be able to fully communicate to others. Let the flood-gates of truth once be opened and come in upon you as they have upon me, and you will be overwhelmed by the mass of their detail no less than by the vigor of their mass. My great want, therefore, for the purpose of more fully arriving at these facts and obtaining ever higher results is assistance and coöperation. I wish it to be distinctly understood, however, that I do not mean this in a personal sense—far from it; but in the interest and the promotion of science, as everybody wanting to make original observations must pursue these studies for himself and by himself.

Why such a course has not been heretofore pursued by others I am at a loss to understand, except from the fact that it takes an unusual amount of perseverance to reach the first results. Though *all* persons may not be able to personally obtain satisfactory results, *all* may be *benefited* by the results obtained by those qualified to successfully carry on a course of observations by means of introspection. The world at large will always have to be satisfied with being simply the beneficiary of scientific research; more especially of research in matters spiritual or psychical. From facts thus obtained rules may be deduced, which, translated into "physical forms," may become the property of all. In this manner numerous observations I have made have already assumed a practical shape; but I have not as yet been able to devote the necessary time to them to produce a system which may be used for general instruction.

Meanwhile I do sincerely hope that others will take hold of these matters in all seriousness, and assist me in arriving at these practical physical forms, which I trust, in fact *know* beyond the shadow of a doubt, will be fruitful of the most beneficent results in the teaching of the deaf, of singing and elocution, of pure vocal utterance in speaking; in curing stammering and other chronic faulty or deficient utterance; besides numerous other matters of equal importance not in immediate connection with vocal utterance.

That these matters must be and are of the greatest importance to the medical student goes without saying. It is to be hoped that they may lead to a more rational treatment of our frail and often ailing bodies. I say "bodies" because this is the common phrase. Yet how false this is, every true physician is but too conscious of. Our ailments cannot be successfully treated from a mere physical standpoint. The question of life is not a mechanical one; it is spiritual beyond anything else, the spirit being the motive power giving life to the otherwise inert physical body. Yet the only endeavor of the physician has always been to cure the "machine," to set its mechanism right again when it is out of order, simply because he has not been able to get at the spiritual motive power which propels it.

I have been trying to get at this motive power, and to some extent have been successful in so doing. Besides, the *body* never suffers. Its ailments make the soul suffer; while the ailments of the soul have a comparatively less injurious effect upon the body. The body is the habitation of the soul. The soul dwells in its *every* part. As long as this habitation is habitable the soul continues to dwell therein. When it becomes uninhabitable the soul departs, never to return. Hence a body, never so frail and ailing, will continue to live as long as a vital part is not affected, that is, a part the soul *requires* for its habitation and cannot do without. Close such part to the indwelling of the soul, prevent material and spiritual factors from joining hands therein, and the spirit departs. Once departed it can never be made to return. Hence a body in the full vigor of health, after having been immersed in water sufficiently long to have any one vital avenue positively closed against the indwelling of the soul, cannot be resuscitated. As long as the soul clings to it, however, with never so feeble a grasp, it may come to life again, in the same manner that a flame nearly extinguished may be fanned to life again.

For me to *fully* describe my mode of proceeding in arriving at these matters would be equal to an attempt at crowding into a few paragraphs *all* I have gone through within something like forty years, more or less, of observation.

MAKING PARTS RIGID

I have already stated that I was originally led into making these investigations through my simple desire of getting rid of my *German* mode of expression in speaking the English language. Being determined to find out where the trouble was which prevented me from producing pure English sounds while I experienced no difficulty in producing pure German sounds, I pursued vocal sounds, through numerous phases, to their original sources. The endeavor to arrive at the true nature of vocal sounds through autology and by means of "introspection" has, no doubt, been made by thousands before me. The reason they were not more successful must be attributed to the simple fact that such persons have been lacking in perseverance. It is one of the most misleading endeavors one can pursue.

In the beginning I came to what I considered a *positive* result perhaps for the hundredth time, but to think I was on the wrong track the one hundred and first time. I would then, perhaps, finally determine that the first result arrived at, after all, was the correct one. In this manner I have in the course of time arrived at positive conclusions, which have been the basis of all my investigations, and are undoubtedly correct, as they have yielded up one result after another and have never proven false. For this, relatively speaking, "perfect insight" I have waited, before saying anything more at all, since my previous (preliminary) publication. To these conclusions I owe my present trust and confidence, and the "boldness and temerity," as some may say, in making such "startling declarations" in the face of the accumulated wisdom of the science of this and of past ages. Yet I am tired unto death of prevarication and of time-serving, and will say what I consider to be the truth, no matter what may be the consequence.

Any one singing a false note or mispronouncing a foreign word or sound, yet knowing what the right note, word, or sound is and should be, can do the same thing, and by perseverance finally find what he has been looking for and pronounce such note, word, or sound in its entire purity. This will put him on the track to the production of *all* pure notes or sounds. To accomplish this, he must persistently watch one result after another.

My mode of proceeding has been largely in making parts *rigid*, and then observing the consequences. In pursuing this course for some time, you will finally attain such a mastery therein that you will be able to make almost any vessel, muscle, sinew, membrane, tissue, etc., or any *part* thereof, rigid. This is done for the purpose of neutralizing parts which partake in the production of sounds, and will enable you to closely watch cause and effect in your natural, as well as artistic, course of breathing and sound production. *Having two languages at my command, I was startled to find that cause and effect in both were totally different from each other.* This gave me the original cue to all my observations.

In place of sounds, others may pursue odor, taste, feeling, motion, hearing, etc., to their original sources, and make similar observations. In so doing they will find that *all phenomena, the products of our faculties, abilities, or gifts, originally proceed from the same or similar sources; that there is a homogeneity of proceeding, mainly consisting in various modes of breathing, in the production of them all; the end organs of our senses or gifts finally determining definite special results.*

For vocal utterance, we draw our inspiration for various results to be attained, from the air, and breathe in a different mode for every special performance. These modes of breathing, though the same for all persons in a general sense and leading through the same channels, in a more restricted sense are different for every nationality.

There is no "danger" connected with these pursuits, in spite of Mr. Heidenhain's fears; which fact is due to the duality of the nature of each and all our various faculties, there being a safety-valve always at the other end in the shape of the negative factor. The only danger I have discovered was in connection with the "streams of life," which do not permit tampering with without penalty. As these exist independent of our ordinary mode of breathing, they are not apt to be interfered with by any neophyte in the pursuits now under consideration. Of these powerful streams, of which no notice has ever been taken by any one, though ceaselessly streaming into and out of our system while life lasts, I shall take occasion to speak later on.

EXTIRPATION

To make a part "rigid" is equal to the "extirpation" of such part. While it is in a state of rigidity, it ceases to take part in any action whatsoever; it is inert and the same as if it had ceased to exist. What advantage, then, let me ask, is there in extirpating parts in animals, when we can, by making parts rigid, directly extirpate such parts in ourselves? We can in this manner suppress the action of any muscle, or the participation of any vessel, or part of such vessel, in any act, by the simple exercise of our volition. I find no difficulty in thus "extirpating" any such part from myself for the time being, and then observing the consequences. I can take hold of the innermost part of myself, so to say, and take it *out of myself.* In regard to vocal utterance, these consequences are positive and direct. That these operations must be very *carefully* conducted in connection with *vital* parts goes without saying. The action of muscles participating in the production of vocal utterance, however, or in the act of breathing, except the muscles of the heart, can be suppressed without danger. I am thus in a position to modify extirpation of parts to any extent, almost, I desire. I can add to and detract therefrom at will, and can shift the act of extirpation from the anterior part of a vessel to its posterior, or from its superior to its inferior, or vice versa, now making one side rigid, then the other, now one end, and then the other; or take hold of its centre and leave the other parts free, or suppress its circumference and leave the centre free. There is scarcely a limit to the action of my will in handling my subject. All this while, my feelings, my intelligence, my mind, take in every phase of these proceedings, and enable me to give a correct account of the results I have been observing.

This discovery—for a discovery it must be, as I can find no account of any similar proceeding ever having been carried on—should, and I hope will, put an end to vivisection, when it is resorted to for the purpose of learning anything whatever in respect to the action and the process of life. By this proceeding I have more or less successfully observed the acts of breathing, of vocal utterance, motion and locomotion, hearing, seeing, and thinking.

I beg leave to here insert without comment the following clipping from the press:

The following extracts are from a lecture on "Vivisection in Relation to Medical Science," delivered by Edward Berdoe, M. R. C. S., etc., at Cambridge. Lovers of animals may be glad to know how the medical fraternity amuse themselves:

"You may open the abdomens of living cats, guinea-pigs, and rabbits, and apply irritating chemicals to their exposed intestines, causing what you are pleased to term 'peculiar rhythmic movements' and 'circus movements,' but what the unlearned would call violent spasms and convulsions, as was done by Dr. Batten and Mr. Bokenham, at St. Bartholomew's Hospital, last year. You may dissect out the kidneys of living dogs and cats which you have first paralyzed by curare—the 'hellish oorali' of Lord Tennyson's poem, so called because the animal's sufferings are intensified by its use, and it is unable to move a limb, or to bite, scratch, howl, or otherwise interfere with the operator's comfort. You may do this, as was done by Dr. John Rose Bradford, at University College, London. You may infect ninety cats with cholera poison, and bake numbers of them alive, as did Dr. Lander Brunton. You may inoculate the eyes of rabbits and guinea-pigs with the material of tubercle, fix glass balls filled with croton oil—a horribly irritating drug—and stitch them into the muscles of the backs of rabbits, then crush them amongst their tissues, as did Dr. Watson Cheyne, at King's College, London. You may slice, plough, burn, and pick away the brains of monkeys and dogs, as did Dr. Ferrier. You may slowly starve to death animals whose vagi nerves have been cut and stimulated by electricity, as was done by Dr. Gaskell, of this University, in 1878. You may cut out the spleens and livers from living rabbits, pigeons, and ducks, as was done by Dr. William Hunter, of St. John's College, Cambridge, in 1888, or do a thousand other acts which in a coster-monger or a farm laborer would be termed and dealt with as acts of atrocious cruelty, punishable by imprisonment. But you have not learned the cure for a single malady which afflicts the human body."

THE MOVEMENTS OF THE TONGUE

There is another mode of proceeding by which satisfactory results can be obtained, and which was the only one I resorted to in the beginning and for many years afterwards; namely, the watching of the movements of the tongue.

The muscle of the tongue, for vocal utterance, is the most important in our organization. It appears to me, in fact, as if in its tip there were a concentration of all the threads which control our existence; and that it is, therefore, representative of an epitome of our entire being. As all sciences, in a general, though in some instances perhaps somewhat remote, sense, centre in the science of life, so do the controlling elements in our composition centre in the tip of the tongue. If it were possible to analyze it spiritually as well as physically, we would obtain a compendium of knowledge far in advance of any there is in existence in the world at the present time. Still, it must be admitted that this would, to some extent, depend upon *whose* tongue's tip was submitted to such analyzation. The fact of the tip of the tongue being removed by surgical operation without serious effect upon the mental condition of the individual does not greatly affect my assertion. In that case the concentration must have taken place at the tongue's new tip or end.

The tongue's tip, with as infallible correctness as the magnetic needle points towards the north pole, indicates the exact spot whence sounds come, or should come, to appear on the surface in a clear and undefiled manner. The tongue's tip, for English vowel sounds, does not touch any part of the oral cavity. It is constantly changing its position, however, and for every vowel sound, or shade of a vowel sound, points in the direction of or *approaches* the spot whence a sound comes, or should come.

To ascertain such spot with exactitude, it is but necessary to *extend* the tongue's tip until it reaches the wall of the oral cavity during or, still better, immediately after the utterance of a vocal sound. Upon reaching that spot the tongue may continue in the same position of contact and the sound can still be uttered with entire purity. Change this point of contact, however, but in the least, and such sound will at once cease to come to the surface. Yet, while *apparently* a sound comes from the direction in which the tip of the tongue points, this is not really the case. In pointing in a given direction, the tongue opens up the channels of the œsophagus and the trachea in a special manner for the proper emission of a given sound, beneath as well as above, and to the left as well as to the right of its radix. In changing the tongue's position but in the least, these channels will open in a different direction, which may then be the proper medium for the emission of another sound, but not for the one under consideration.

The general mode in which the radix of the tongue turns upon its axis is the direct and fundamental cause productive of the various languages of the world; such general mode necessitating special movements of the tongue for the production of the sounds of any special language. Regarding the proper emission of consonant sounds every one knows that the same depends upon the particular spot of contact of the tongue's tip with parts of the oral cavity. As a matter of fact, such point of contact also opens, the same as with vowel sounds, the tubes of the trachea and œsophagus at the tongue's radix in the proper manner for the emission of a given stream of air for the production of such consonant sounds.

Every imaginable opprobrious epithet has been by singers bestowed upon the tongue. "This obstreperous muscle which is always in the way," says one. "This troublesome member will persist in going up when you want it to remain down"; "intractable," "contrary," "obstinate," "wilful," "ungovernable," "stubborn." All these expressions have been used by writers on the voice in connection with the tongue, simply because it would not yield to unreasonable and unnatural demands made upon it; the tongue, being a free agent, persisting in its natural rights—as much so as any independent democratic citizen persists in his.

My observations having been made in connection with a foreign language, I had a better opportunity for watching my tongue's movements than I would have had had I attempted to watch them in connection with my native tongue; the movements of the tongue in connection with the latter being so rapid and involuntary that it becomes exceedingly difficult to make any observations at all. It was like having this foreign (English) tongue exist independently alongside of my own, my intelligence watching it, and guiding it, now here, now there, until it would touch the right spot for the right English sound. Knowing what the right sound was and should be, I never stopped until the same came to the surface.

In trying to find my way in this foreign (English) territory of the oral cavity, I might compare my English tongue to the stick in the hands of a blind man, who uses it in place of his eyes to ascertain his whereabouts, so as to enable him to proceed on his way in the right direction. With my "stick" I felt in every direction, till I found I could steer clear of obstacles straight into the channel of the sound I had been seeking. From my German post of observation I was thus enabled to watch the movements of my English tongue in its efforts to find itself "at home" in this foreign territory, while I was at the same time guiding it from one point therein to another.

I want to call especial attention to and reiterate the fact that the exact point whence a sound proceeds, or seems to proceed, can, by extending the tongue's tip, be quite as well (if not better) ascertained, *after* the utterance of a sound, as *during* such utterance; that is *immediately* after the tongue has ceased to vibrate for such sound.

The difference in the movements of the tongue for various languages is one of the most interesting observations to be made in connection with these studies. The German language being the exact opposite, the antipode, to the English, after comprehending the movements of the tongue for the latter, its own movements, that is, the movements of the tongue for German sounds, were not difficult for me to ascertain.

It is an anomaly to apply the works of German writers on the voice to the study of the English language, or to that of any other than the German language; or to apply books written from an English standpoint to the study of any language except the English—the movements of the tongue, and, in sympathy therewith, of countless other muscles, being different for every language.

Whatever the movements of the tongue are for the *spoken* language, they are of an inverse order for *song*. I anticipate in making the following statement, namely, that while speech is of an order which is rapid, direct, anterior, exterior, spontaneous, impulsive, and material, song is of an order which is slow, indirect, posterior, interior, premeditated, contemplative, and spiritual. I will also add this: that, *while speech is of the oral cavity, song is of the pharynx*. In making these remarks and others *in anticipation*, I do so intentionally and for a purpose; not so much in expectation that they will be at once and fully understood, as with a view of setting others thinking on these subjects until I can reach them in due course of time; or, if I should *never* be able to reach them, that the principle, at least, underlying the same, which if the opportunity had been granted me would have been fully sustained, shall not be lost. The reader will notice that I am hurrying over the ground as rapidly as I consistently can, even from my—under the best of circumstances—superficial standpoint, leaving wide gaps to be filled in by others in the course of time.

SIMPLE SOUNDS

Speaking of sounds in making experiments in connection with the movements of the tongue, it is of the first importance that these sounds should be *simple* and not *vocal* or compound. They must be sounds of the same order as we utter in whispering, or such sounds as we are apt to use when learning to speak a foreign tongue. They are the inharmonious sounds of the deaf, and those which distinguish the speech of a foreigner from that of the native-born.

The recognition of these sounds as the *negative parts of speech* has been one of my main accomplishments, and has been of the greatest assistance to me in my investigations.

Things *complete* tell no tales. We must decompose them, reduce them to their elements, if we want to arrive at the truth in matters of science. I have succeeded in doing with things spiritual—vocal sounds—what the chemist is doing with things material. In things complete, as they are shaped by the hand of nature, the elements of which they are composed are mingled in such a dexterous manner, are so happily blended, that they adjust, counterpoise, and complement one another, and thus live with and in one another.

These new forms have been created by the elements of which they are composed, abandoning their separate original forms and now appearing in a new form, as integral parts of an *harmonious* entity. These elements have not only abandoned their form, however, but in most instances have also changed their character; which in their original composition may have been of a *discordant*, violent, and even dangerous nature. Take but the atmospheric air and its elements for an example.

A similar state of affairs exists in connection with the phenomena of the material-spiritual world. While vocal sounds, when properly produced, stand for all that is harmonious and pleasing, their component parts, their positive and negative elements, by themselves, offer features of a contrary nature. They also offer us, the same as elements do to the chemist while making experiments, the opportunity for making an endless number of combinations. Unless you know what *simple* sounds—*i. e.*, negative parts of vocal sounds—are, and know how to produce them, you will scarcely be able to make one class of experiments which I shall offer in great abundance to sustain my arguments.

When I shall reach the subject of vocal sounds proper, I shall more fully explain their exact nature. I will simply say this at present: A simple sound is the product of that hemisphere only to which it properly belongs. A vocal sound is aided and assisted by a complementary sound from the other hemisphere. The more perfect such aid, the more perfect will be its tone. Simple vowel sounds are short, abrupt, the same as consonant sounds when produced all by themselves and without the aid of a vowel sound uttered in conjunction with them.

POSTERIOR SURFACES

In saying, as I have, that introspection is carried on by looking into ourselves with the *inner surface of our eyes*, I meant to say, in the first instance, that we must exclude all exterior vision, and then attempt to locate and follow up the course of events going on within us. While in this state we are strictly reduced to our personal and individual existence. In thus "watching," the function of our eyes, instead of being used for external material observation, is reversed; their function now being to observe internally and spiritually.

In connection with sounds, you will not only "in your mind's eye" *see* the places where they originate, and *feel* the course they are taking, but you will actually, functionally (in the mode of spiritually seeing and feeling), "see" and "feel" them. This vision and this feeling is far from being perfect, however,—not being accustomed to thus seeing and feeling,—but it may, when continuously exercised, become so in the course of time. While in this state, besides seeing the places interiorly, you may also see them exteriorly, by reflection as it were, and in a reverse order, "as in a looking-glass," in which case it is still an interior vision reflected exteriorly. As a matter of fact, I not only believe, but positively *know*, that *every exterior functional surface has a corresponding posterior one.*

Whenever a thing is brought *home* to us, either through our organs of seeing, hearing, feeling, smelling, or tasting, the outer surface of such respective organ constitutes the positive factor for such action, while its inner surface constitutes the negative factor thereof. Whenever the outer world is excluded, however, as during thought, introspection, and in our sleep, the inner surface of any of these organs becomes the positive, and the outer surface the negative, factor. In thus saying, "I see with the inner surface of my eyes," I do not mean this figuratively only, but literally, functionally, as well; as I could not see these places and locate them internally nor could I see any subject or object with "my mind's eye," if the faculty of seeing were not actually given to the posterior surface of the eye.

This will become clear when you consider that you will altogether fail to see internally when you attempt to use the *anterior* surface of your eye for the purpose of *internal* vision. Thus, the phenomena of vision which accompany thought or dreams, during sleep as well as in our waking moments, are not merely spiritual, but, in the sense of internal functional vision, are also material, so to say. *All* thought, in fact, is more or less of this same nature. We use the posterior surfaces of our organs of sense more frequently, in consequence, than we do their corresponding anterior surfaces. Physiologists will say there is no such a thing as an inner surface of the eye capable of seeing. This does not alter the fact that I actually, functionally, see with the posterior surface of my eyes, and that everybody else does the same thing.

I shall, in connection with vocal utterance, have occasion to call attention to numerous divisions of as positive a character as a wall of living tissue, of which there is not a trace to be seen by external vision; these divisions being channels, constantly used in one and the same direction, some for ingoing, others for outgoing streams of air and sounds. Of these channels, also, being invisible to the outer surface of the eye, science has never taken any notice. These invisible agencies are connecting links, mediating between cause and result, in connection with material-spiritual or spiritual-material phenomena of whatsoever nature brought to our consciousness. Hence the inability of science, in its ignorance of these agencies, to reconcile the one with the other by the aid of such material only as has been heretofore at its disposal. We may *see* proceedings going on which are mediating between cause and effect, by the assistance of the inner surface of our eyes. They disappear altogether, as well as any other "vision," upon an attempt being made at seeing them with the external surface of our eyes. Yet we may see inwardly with our eyes open, as we do when absent-minded, etc.

If we could invent a microscope by the aid of which we could look into ourselves in a *spiritual* sense, that is, through posterior surfaces, *all* the secret springs of our nature might be revealed to us. This ability to become cognizant of physiologico-psychological processes by the aid of the inner surfaces of our organs of sense, reveals a peculiar functional exercise of their faculties. In matters of memory they are not intended to aid in conveying to our consciousness impressions made at the *present*, but those made at a previous time. These impressions having been made on the soft tablets of our brain, either during our individual existence or that of our progenitors, and transmitted to us by dint of heredity, are brought to our consciousness by the aid of these inner surfaces, *phonographically*. They are awakened by association; and that organ of sense by the aid of whose anterior surface they were first received and *recorded*, now reawakens them by the aid of its posterior surface. Visions, consequently, are reflections made on the inner surface of the eyes, from impressions previously made upon the brain, in a similar manner to that by which sounds come forth from a phonograph. They could not assume shape if they were not thus reflected. It is owing to the nature of these reflections that they are more fleeting and evanescent than those made by the objects themselves upon the external surface of the eyes.

The anterior and posterior surfaces of all organs, by whose aid we exercise our faculties, which surfaces represent their poles and dual factors, the positive and the negative, the material and the spiritual, change places in conformity with whether an object is impressed upon them exteriorly or interiorly, in the present or the past, directly or indirectly, physically or spiritually. Things which are brought to our consciousness from the exterior world and in a direct manner—through our senses—may be said to be of a *material* nature; while those which come to us indirectly—through our inner consciousness—may be said to be of *spiritual* origin. The clearness of our visions naturally depends upon the clearness of the impression still remaining upon the tablets of the brain. The more stirring the event in the first instance, the deeper and more lasting, of course, the impression. All this, however, does not throw any light upon the process of abstract thought; nor am I in a position to aid in so doing. Yet it appears to me to be a sister proceeding; and that a nearer approach to an explanation of those more material phenomena may finally assist in arriving at an explanation of the causes of these more recondite and apparently purely spiritual phenomena.

The correctness of the preceding remarks will become more apparent when we substitute for the faculty of seeing, that of hearing. We hear the voice of another person through the *anterior* part of our ear, *entering*, as it does, from *without*. We hear our own voice through the *posterior* part of our ear, *going out*, as it does, from *within*. No matter how low we may speak, we can always hear our own voice, though inaudible to others; and we can still distinctly hear it at such time, even when we fail to hear a low, though in fact relatively much louder, tone proceeding from the voice of another person. A ventriloquist, on the other hand, with whom these relations are reversed, hears his own voice reflected from without, inwardly, while, if he continues in the same condition while listening to another person's voice, he will hear the latter from within, outwardly.

For the purpose of testing the correctness of these observations, please pay attention to the following: In listening to the sounds of another person's speech, you will have no difficulty in noticing that they stream into your ear from without, inwardly.

Now, substitute for this other person's voice the sounds of your own voice, *and continue to listen to the same in precisely the same manner in which you did to those of this other person*; that is, let them flow into your ear from without, inwardly. The result will be *that you will not only not hear the sounds of your own voice, but that these sounds themselves will become paralyzed, that you will not be able to produce any sound whatever.*

The cause is obvious. You attempt to listen to negative sounds with the side of your ear still tuned negatively; while, ordinarily, when we cease to listen and commence to speak, *all* poles are reversed. Spoken sounds are positive in relation to the speaker, but negative in relation to the person listening to the same. In consequence, the producer hears them with the negative (inner) part of his ear, the receiver, or listener, hears them with the positive (exterior) part of his ear.

I copy the following from an article in the *Philadelphia Sunday Press*:

"A curious fact in regard to the effect of explosions upon the drumhead, is that this tissue, though generally blown in, is sometimes blown out. Just what causes the latter result has not yet been fully explained."

In this instance, I presume, the person's ear was tuned to listen interiorly, and the effect of the explosion, which, in relation to him, was of a negative nature, took effect on the positive, the posterior, side of his ear. This person was not in expectancy of the explosion, but it came on unawares, of a sudden, while he was in a state of contemplation.

In connection with the eye, our inner consciousness acts as a "rein" upon the outer, drawing back in case of danger, checking our progress when suddenly coming upon a precipice, and *regulating our steps* to circumvent it, but without coming to a stop, when seeing an obstacle in our way from a distance. The "rein" in such an instance reverses the poles of the eyes—the positive becomes negative and the negative positive; that is to say, in our usual mode of seeing, while walking, the exterior surface of the eye is positive, the interior negative; but when there is danger ahead and we are warned to be cautious, the exterior becomes negative and the interior positive; the activity now being exercised by the latter, the passivity by the former. The action of the "rein," however, is not direct, but crosswise; that is to say, the posterior surface of the left eye is in correspondence with the anterior of the right, and vice versa, in conformity with the "impulse" emanating from either the one or the other, while the anterior surface of the left eye is in correspondence with the posterior of the right, and vice versa.

The knowledge of the reversion of the functional exercise of our organs of sense is of signal importance in connection with motion and vocal utterance, which always go hand in hand; every utterance being accompanied by a motion, though not always visible to the eye. In truly artistic delivery these motions are brought to the highest perfection; and visibly, though often in great moderation, accompany *every* inflection of the voice.

To be able to see a thing at all, we must be in a relatively proper position with the object to be seen; we must be on the same plane with it. We must also have light, not only for the latter, but by reflection therefrom also for ourselves. In addition we must have the inner light enabling us to comprehend what we have seen. I contend that for the study of spiritual-material as well as material-spiritual phenomena, such light has always been wanting for the thing to be seen, as well as for the orb to see and consequently for the spirit to comprehend. In attempting to comprehend, and to explain appearances, physiologically, we have been looking in our exterior world, where we cannot, in place of our interior world, where we might be able to see and to observe. We have been using the outer surface of our eye instead of the inner, with which to see spiritual things. The thing to be seen and the orb with which to see were not on the same "plane." It was impossible to perform the act of *spiritually* seeing. The proper light once obtained, it has not only illumined for me the things to be seen, but also my capacity for seeing and comprehending them. Roentgen has taught us the method of seeing material things through opaque bodies. I have learned to recognize spiritual phenomena in opaque bodies, created, as they are, by a combination of spiritual and material factors. While I have made use of this gift for a special study—that of vocal utterance—I incline to think that it may be made use of for the study of not only all the various material-spiritual phenomena to be observed in the nature of organic bodies in general and man's in particular, but also of our relations with the unseen and unknown world and its forces, in which our essence has its being, whence it comes, and to which it returns. In minutely explaining my mode of proceeding, it is also my special desire to rob it of any appearance of "supernaturalness" some persons might be inclined to invest it with. Though I cannot explain many things connected with the voice from an entirely naturalistic standpoint, I think they are all explainable if the proper amount of study and observation be given to them. This, as a matter of course, does not, however, include the operations of the mind proper, which are governed by laws beyond any human understanding.

INSPIRATION—EXPIRATION

The entire mechanism of our being, more especially that of our faculties and functions, is primarily excited through openings into which air is inspired, from which air is expired. These openings are connected with channels and vessels which are passive or negative during inspiration; active or positive during expiration. Thus the multiform streams of air introduced into our system communicate with parts thereof, which, by their construction and intercommunication with others, are specially adapted for the exercise of any special faculty or function. Our will directs these streams of air to flow into their proper channels (and they automatically obey) for the guidance of our steps in a certain direction, for the production of a given sound, the recognition of a given sight, the sensation of a peculiar odor, taste, or feeling, or the excitation of a passion, a compassion, or any other sensation, feeling, or thought whatsoever. These streams of air, therefore, are of an order as multiform as the complex web of our material and spiritual existence, and are introduced through thousands of different channels and in thousands of different ways.

To confine our mode of physical and spiritual existence to a single stream of air introduced into the oral cavity, or the nostrils, and thence into the lungs, appears to me to be as primitive a proceeding and as narrow a view as can possibly be taken of one of the greatest subjects our understanding is called upon to deal with. In place of that, I have positive proof that the streams of air which flow into these openings are of the most multiform nature; every sight, odor, taste, touch, and every sound, and fraction of a sound even, calling for a special stream of air which no other stream can furnish or supply. Besides the oral cavity and the nostrils, the eyes, ears, and every additional opening, down to an almost invisible pore or capillary vessel, are recipients of special streams intended for special purposes. *We breathe through the soles of our feet and the palms of our hands, as well as through the skull of our heads. The closer we guard our body against the influence of the air, by means of unnaturally close-woven and air-tight clothing, the less capable we become of exercising our natural faculties and functions.*

To this subject I shall devote time and attention at some future period, more especially in connection with vocal utterance, as it has everything to do with the production of sounds, which proceed in part from within, outwardly, and in part from without, inwardly. In so doing, positive becomes negative and negative positive; inspiration and expiration equalize each other, and thus a continuous flow of speech becomes possible, while if the flow were continuously in one and the same direction it would soon come to an absolute stop.

It is this that science has done for us: It has clogged up all these natural avenues to our existence by teaching that we breathe through the trachea alone, in consequence of the muscle of the diaphragm forming an air-tight partition between the upper and lower compartments of our bodies; being ignorant of the fact of that other great tube of the œsophagus, also opening into the oral cavity, performing the same functions for the abdomen which the trachea does for the thorax. In place of all these millions of openings through which we inspire and expire, science teaches that we breathe through a single tube, into and out of an *air-tight sack*,—a mechanically impossible proceeding. By some ill-defined process, air is supposed to find its way into the thorax and out again after depositing its oxygen in the blood-vessels. Meanwhile, the balance of our body is left to shift for itself, not the slightest particle of fresh food ever finding its way into any portion thereof, except indirectly through the blood-vessels. To my simple and untaught understanding it appears that if such a state of affairs really existed—no matter how rapid the circulation of the blood—the entire hemisphere of the abdomen would be given over to putrefaction in an exceedingly short space of time.

Breathing, however, as we do, through the œsophagus, in like measure with the trachea, and through every other opening in our epidermis in addition, our body is constantly, uninterruptedly, permeated with fresh air in its every avenue, vessel, capillary tube, cell, etc., which sustains us by its life-giving qualities, and takes away with it the constantly accumulating refuse.

The muscle of the diaphragm has been the air-tight door to the cell of the condemned, whose portal has been guarded by ignorance and every oppression, suppression, fear, superstition, anxiety, bigotry, narrowness, prejudice, etc., that the human mind is capable of. It has given us over to self-accusation as a natural and vital element. It has shut us up into the narrowest limits, and kept us from communing with the universe and the spirit of the universe. It has excluded from us the grace, the beauty, the light, the liberty, the eternity of the *spirit*, and prevented us from recognizing ourselves as integral parts of the universe and of the causes which sustain it and sustain us. It has prevented us from communing with them as free agents *in our own name and by our own right*, without interference or the intercession of any person or agency whatsoever, in the past or the present.

Have I placed too great a value on the discovery of the "voice of the œsophagus"?

I feel convinced that the further exposition of my observations will justify me in all I have said.

DIAPHRAGMS

As the trunk has its diaphragm, dividing thorax and abdomen, so do all dual hemispheres representing a faculty or function have their diaphragms, performing duties of an analogous nature. *Every* opening, in fact, has its diaphragm. Where there is none visible, it is formed by contraction, whenever needed, and but for the time being. All these various diaphragms, more particularly the one specially bearing that name, are of the greatest importance in connection with vocal utterance,—the sounds of the vessels of the abdomen being produced by an expansion of the thorax and consequent contraction of the abdomen, those of the vessels of the thorax by an expansion of the abdomen and a consequent contraction of the thorax.

For the purposes of vocal utterance, inspiration into the thorax produces an expiration from the abdomen by way of the œsophagus, accompanied by vocal sound, while an inspiration into the abdomen produces an expiration from the thorax by way of the trachea, accompanied by vocal sound; the special *mode* of inspiration regulating the special sound to be produced.

This proceeding has reference to outgoing sounds only. For ingoing sounds the opposite proceeding takes place; an expiration from the thorax producing an inspiration into the abdomen, and an expiration from the abdomen an inspiration into the thorax, both accompanied by sound. Every original inspiration into thorax or abdomen, of course, must have been preceded by an expiration from these parts, while every original expiration must have been preceded by an inspiration into the same. The utterance of every sound, therefore, requires at least three movements on the part of the respiratory organs. But for the action of the diaphragm, such sounds could not be produced.

All these various diaphragms fall or recede for inspiration, rise or advance for expiration; the function of a diaphragm being exercised in conformity with the manner in which it is approached. This may be done by way of the œsophagus or the trachea, *i. e.*, from the side of the hemisphere of the abdomen, or from that of the thorax. The outward movement of the abdomen during respiration, therefore, is not caused by a pressure brought to bear on its contents by the diaphragm, but it advances and recedes in conformity with a direct process of inspiration and expiration by way of the œsophagus and the trachea; the œsophagus and trachea sustaining each other and acting reciprocally and in conjunction. This presumed pressing forward and subsequent receding of the entrails, in consequence of the descent and ascent of the diaphragm, presents a spectacle as repugnant as it is impossible of execution; the extension of the abdomen, more particularly in connection with special sounds, being so great that no pressure whatever brought to bear upon the entrails could possibly produce it.

In place of this theory, now so generally entertained, the simple fact obtains that the diaphragm descends in consequence of an influx of air into and subsequent expansion of the thorax, causing a contraction of the abdomen and an efflux of air from the same; that it ascends in consequence of an influx of air into and expansion of the abdomen, causing a contraction of the thorax and an efflux of air from the same.

IMPRESSION AND EXPRESSION

All vocal expression is but an echo, the echo of a thought. Thought *must* precede vocal expression. It is not possible to produce a vocal sound, not the simplest, without thought. There is no such thing as a voice *ipso facto*, no more than there is music in a musical instrument unless it is called forth by the hand of the player. Try it. Come upon a sound suddenly, around the corner, as it were, and then express it. Do not give it a moment's time for its development; that is, do not give thought time to mould a form for it, but try to utter it in embryo, so to say, the very moment you think of it, and you will not be able to do it. You will not produce any sound whatever.

It is as necessary to form a mould for a sound as it is for any shaped and moulded material article. Out of this mould it comes forth in conformity with the form we have given it: harsh, abrupt, discordant—rhythmical, beautiful, soulful. Such as the thought is, so will be the expression. In ordinary conversation this proceeding is automatic and mechanical, in elocution or song more or less volitional and artistic. That is to say, for ordinary speech it acts automatically, for artistic utterance it acts designedly. Materially, the mould is convex, shut, for ingoing; concave, open, for outgoing sounds. It expands for the former, it contracts for the latter. Vocal sounds are a product of matter as well as mind; the act itself which produces them being a connecting link between matter and mind. The soul calls on the body to aid it in giving form to its desires and intentions; the body instantly obeys and assumes the form from which the expected sound or action is to arise.

No matter how great a soul may be, unless it can give form and consequent utterance to its greatness, it will be helpless, far more so than the simplest soul capable of giving expression to its simplicity. Confined to our own limits, like the congenital deaf, our faculties become dwarfed and useless. We do not know ourselves, do not know our own souls. We must expand, go out into the world and take it in, if we want to grow and give our faculties a chance to develop.

The greater our horizon, the more we can take in, the more we can give out. Our soul is scarcely ours when enchained; the greater its liberty, the more it belongs to us. Hence our just pity for the congenital deaf, and our desire to assist them in their efforts at expression. Those among them who are being, or have been, tutored, receive their impressions through their eyes in the form assumed by the speaker's mouth; the eye assuming the function of the ear. The form assumed by their teacher's mouth, however, not being perfect, a perfect impression cannot be made. Hence the expression of the deaf is in conformity with the impression they have obtained: mechanical, material, soulless. The exterior lines of the mouth of the teacher, or any other speaker's from which the deaf draw their inspiration, are those of the material side of the medal. Failing to see the reverse side thereof, namely, the interior of the mouth, which is its spiritual side, the lines of the latter make no impression upon them. These fine lines on the interior side of the speaker's mouth, representing the rhythm, the soul of the voice, not being seen, fail to make that impression from which alone a soulful expression could arise.

That an *impression* may be made through the eye will scarcely require a defense, in view of the fact that in reading aloud or in singing from notes the *entire* impression is made through the eye. The reader or singer, knowing the *value* of every sound, is impressed by the sight of a letter or a note as he would be by the sound itself. Not so with the congenital deaf, who, being ignorant of such value, cannot reproduce it. Nor will it be contended, I suppose, that the deaf knowingly, designedly, or volitionally attempt to imitate the forms assumed by the teacher's mouth, but it will be admitted that this is done spontaneously, and that vocal sounds with them arise from this imperfect mechanism, thus involuntarily reproduced.

With the congenital deaf, with persons attempting to speak a foreign language, etc., the material form, as well as the spiritual impetus, being imperfect, the expression will be in conformity therewith. In how far and in what manner these investigations may become helpful to the deaf will be a matter for the not distant future to develop. That they will eventually become of the greatest aid to them I have every reason to believe. Those who have made a study of matters of this kind understand the difficulties surrounding the same. These difficulties are increased manifold where the ear of the scholar absolutely refuses to come to his own and his teacher's aid.

There are forms in which vocal sounds move, well defined and capable of material representation, which are not fully expressed by the shape of the teacher's mouth, nor are they thus expressed by impressions taken by the aid of the camera. Regarding the latter, it is necessary to note that photographic representations of vocal sounds are the result of the combined action of the voice of the œsophagus and of that of the trachea, of material and immaterial factors. Just in how far the latter are capable of being thus represented must, as yet, remain a matter of conjecture.

An attempt at reconciling photographic representations of vocal sounds with the oscillations of the vocal cords is, at most, a one-sided proceeding. To arrive at any correct conclusion at all, it would be necessary to take the vibrations of the "vocal lip" and the frænum into equal consideration.

Regarding our capacity for improving the natural physical and psychical capabilities of the musical instrument of the voice, that depends upon the manner in which we play upon it. As it yields to the slightest pressure of the air, either for good or for evil, we must, above all things, learn how to guide the tip of our tongue in touching its aërial strings or keys, which are far more sensitive than those of any instrument ever produced by the hand of man. It takes years to attain a mastery over the simplest musical instrument; yet it is often expected that the instrument of the voice should yield to the most careless efforts made in the most wilful and indiscriminate manner.

The *thought* of a sound, after *producing* an impression, *guides* the tongue in *releasing* such impression. Unless the tongue touches or moves towards the exact spot which will effect such release, the expression or the sound will not be forthcoming. That the impression, as well as its release, should be properly made, it is necessary to *think* of the sound which is to be produced, in the most precise and correct manner. I cannot sufficiently impress upon the reader's mind the importance this simple lesson conveys. If he will shape his manner of vocal utterance, especially his mode of singing, in conformity therewith, he will be able to improve his voice to a far greater extent than he would by following any or all of the realistic methods now in vogue. This *thinking* of the correct sound must be carried on for the *next* syllable during the *production* of the previous one; and care must be taken not to think of more than one syllable at one and the same time. Unless this is done, no pure sound will ever be produced, the impression made by thinking of a second or third syllable overlapping that for the next following; thus producing a muddle and a discord. Rhythm being the basis for all perfect vocal utterance, a rhythmic impression must be made in order to obtain a rhythmic expression. This cannot be done when the former is not preserved in its entire purity until it is released.

All of us, either during our ordinary speech or during our efforts at artistic expression, are guided by the process just described; unknowingly, unwittingly, properly or improperly, for good or for evil, pursuing this same course. I cannot enter upon these matters to any greater extent at this time, as it will be necessary to first treat of other matters with which they are intimately connected.

THE PHONOGRAPH

In trying the experiment of coming upon a sound unawares, simply endeavor to divest yourself of all thought, and then suddenly, without any preparation whatever, say "a," or "b," or "it," or any word you wish, and you will not be able to produce such sound or sounds—or, in fact, any sound whatsoever. Or, you may get some one to, of a sudden, produce sounds embodied in letters before your eyes; and you will find you will be unable to utter them instantly. While you cannot thus produce a vocal sound, or vocal sounds embodied in words, you can produce *simple* sounds without preparation. As they belong to but one hemisphere, and are consequently not the product of a compound impression, they may be uttered the very moment we think of them. While they are being uttered, our organs of speech are "shut," far more so than they are for *vocal* sounds.

Consonant sounds cannot be uttered "vocally" without a vowel sound. When they appear in a syllable their *accompanying* vowel sound carries them and permeates them. When they appear singly we add a vowel sound to them. We say: "ar," "be," "en," "ka," etc.; unless we do so we cannot pronounce them. Without such accompanying vowel sound they would be inert.

"Simple" *consonant* sounds are unaccompanied, not "leavened," by a vowel sound. "Simple" *vowel* sounds, on the other hand, are unaccompanied by the element which constitutes consonant sounds; while "vocal" *vowel* sounds *are* accompanied thereby.

The word "surd," used in connection with non-vocal sounds, does not express the meaning of what I call "simple" sounds, as all sounds may be either "vocal" or "simple," while "surd" applies only to special sounds.

The necessity of making an impression for vocal utterance also prevails in connection with motion. You cannot lift your right foot or your left arm, or make any given motion whatever, the very moment you think of making it. It requires some preparation; though you may lift *part* of a limb without preparation. A part of a limb in this sense may be compared to a *simple*, the entire limb to a *vocal*, sound. The thought must make an impression by expansion or contraction, which, when released, will express the desired motion; no matter whether such motion is made unconsciously or deliberately. It is more difficult to watch this proceeding in connection with sight; the operations of light being so rapid that the expression seems to be simultaneous with the impression.

Contraction and expansion for motion are of the same order as they are for vocal utterance. In fact, both are so closely connected that we cannot utter a sound unless it is accompanied by a motion. In stopping the motion accompanying a sound, we stop our ability of uttering such sound. I shall have occasion to call attention to numerous conditions under which it will be impossible to utter sounds, either separate or connected, by stopping the motion necessary to produce such sounds. It is all due to the fact that we are homogeneous beings, *whose powers are interdependent upon one another.*

The effect of the teacher's *voice* upon his or her scholar's organization is of a *similar* order to that made by *thought* upon the teacher's own organization. That it is not of the *same* order is due to the fact that the organization upon which it is made is but rarely constituted the same, is not as highly organized and developed or "schooled," as the one from which the voice emanated. The impression made by the singing-teacher's *voice* is of the same order as that made upon the deaf by the *features* of their instructor which are representative of his voice. We are living, breathing *phonographs*. Every impression we receive through any of our senses must be made in a material manner before it can have its immaterial expression. We engrave upon living tissue, instead of on rubber or wax.

I repeat that, to obtain a pure sound, the *thought* underlying such sound or sounds must be *purely, clearly defined*. We cannot obtain a clear impression from a seal whose engraving is blurred, or when the sealing-wax is not in a proper condition of softness, or when the hand is not steady which makes the impression. The same conditions prevail with vocal utterance. Thought makes the impression; the æther, passing through its narrowed passages at a rate as swift as thought, creates the sound. The impression is made as *thought* progresses, the expression as *sound* progresses. While the *impression is thoughtful, the expression is thoughtless*. While we think for a sound during the impression, we do not think for it during its expression; *but we think, during the latter, for the next sound*. If this were not the case, consecutive speech would be a matter of impossibility. The artist's thought is embodied in the creation of the model for his statue from which a mould is made. The casting of the statue, equal to its expression, is mechanical, thoughtless.

In this connection the brain is of the same order as the tablets of the phonograph. For ordinary use, however, the lines engraved upon it are evanescent; they disappear again with the sound or thought which releases them. Impressions, however, of a deeper nature remain—some forever. The thought or sounds they represent, the same as the lines on the tablets of the phonograph, are released but for the time being and while such thought and sounds (through association) are recalled to memory. The thought and sounds are evanescent, but the lines which represent them remain for further use, the same as the lines on the tablets of the phonograph and the strings of a musical instrument. If we could read aright the lines which the voice makes on the tablets of the phonograph or on the negative plates of the photographer, we would obtain a correct insight into their character. These studies, when fully developed, may lead to a comprehension of these hieroglyphics, the same as the Greek translation on the Rosetta stone furnished the cue to the comprehension of the hieroglyphics of the Egyptian monuments.

STUTTERING, STAMMERING

What is all this I am writing?

It is an endeavor at giving expression to an impression obtained of a great subject imperfectly understood. The general ideas underlying it all are on the lines of truth, but the contours are evanescent, the lines representing special features ill-defined, while the finer shadings are almost entirely wanting. It is a stuttering, a stammering, in matters my mind is too narrow to grasp, incapable of comprehending in all their bearings, impotent to take in in their ultimate relations. Still, I am doing what I can with such material as nature has placed at my disposal. Thought failing to make a clear impression, my pen, I fear, cannot give a clear expression to it all.

Regarding the subject of stuttering proper, I must still preface it with some remarks of a general nature. The influx and efflux of streams of air into and out of our system, called breathing, is of a very complicated nature. While we designate the same by the general terms of inspiration and expiration, these streams are of as multiform a nature as the ethereal fabrics they are intended to weave, whose weft they form, and whose warp is of a more material nature. Call these fabrics what you please—actions, speech, feelings, passions, fancies, sensations, etc. While these streams form innumerable separate systems, they are all subject to one and the same law—rhythm. The more perfect the rhythm the higher the development and consequent performance.

While we always breathe, or should breathe, in the same rhythmic order (the octave) for the sustenance of life in general, we unconsciously breathe in various other measures for an endless number of other purposes. Our dual nature, and the duality of the manner in which we breathe, as a rule enable us to go through these various performances without a disturbance as to the harmonious character of our existence. It is a great orchestral performance by instruments of various kinds and orders, each performer playing his own notes, specially adapted to his particular part and instrument; yet all coming together in one harmonious *ensemble*. This fact finds expression, clearly defined, in the various measures in which metre and rhythm are clad for poetry and song. The introduction into our system of a rhythmic flow of streams of air for the various purposes of vocal utterance is conditioned upon a rhythmic flow of thought.

To perfectly render a poetical conception by words either spoken or sung, the performer's *mind* must be in accord with the rhythm underlying such conception. In that case only will he breathe and, consequently, speak or sing in the requisite manner for such production. I should have prefaced all this by saying that, in the same manner as inspiration and expiration succeed each other in regular rotation, so do the ordinary measures of long and short (¯˘), or short and long (˘¯), in simple forms of poetry, succeed each other in regular rotation; long (¯), or stress, always standing for expiration, short (˘), or repose, for inspiration. *As a matter of fact, however, inspiration is of longer duration than expiration.*

All other forms are artistic, and are produced by a mode of thinking, and consequent breathing, as variable as the subject may suggest or demand. For ordinary speech, while the rhythm is not of the same order as that for poetry, a rhythmic order of some kind must be, and always is, observed. That the rhythm is not noticeable is due to the fact that, while inspiration and expiration in prose writing and ordinary conversation follow each other in regular rotation, they are not always accompanied by sound. Hence the rhythmic irregularities of speech exist only in appearance and in the inartistic manner in which speech is generally, and prose writing often, produced. A person who speaks and writes his language *well*, speaks and writes it rhythmically, always. Good style is synonymous with correct rhythmical expression, superinduced by correct breathing; rhythmic expression depending entirely upon rhythmic impression, and the latter upon rhythmic thought, accompanied by rhythmic breathing.

To write well (that is, a good style), to speak well (as an orator, actor, or elocutionist), to sing well, it is, above all things, necessary that the performer's mind should be in a state of conformity with the situation which is to be described. His flow of thought, and consequent breathing and mode of expression, will then correspond with the scope, drift, and circumstance underlying his performance. Unless this is the case, the latter will be unsatisfactory, unimpressive, unsympathetic. To prove that for a satisfactory performance this *must* be the case, it will but be necessary to call attention to the fact that under various emotions our mode of breathing undergoes great changes—as under fear, hate, jealousy, indignation, excitement, love, enthusiasm, benevolence, languor, apathy, etc. Our breathing under these different circumstances will, the same as the manner of our expression, undergo various stages of change as to time and measure, as well as to rhythm, emphasis and intonation.

The character and rapidity of the flow of our blood is of the same order as our manner of breathing. It is, in fact, as I expect to prove later on, not only of the same order, but of the same origin and regulated by the same causes. The flow of the blood is not merely of a material order, but of a spiritual one as well. While it is acted upon by the mind it reacts upon the mind.

The thought must be measured and restricted as to time, so as to enable it to make the proper impression and produce a corresponding expression *before* another thought comes along crowding in upon the preceding one and in so doing *blurring* the impression made by the latter before it had been given the time to be expressed. If the necessary time is not granted for an impression to be made and for the expression thereof to obliterate the same, the premature flow of another thought, coming on top of the first, will make a new impression over the previous one, causing confusion and making a clear expression a matter of impossibility. Unless our professor, while standing in front of his blackboard demonstrating before his class, has a sponge in his hand, and before again writing in the same place wipes out that which he had written before, the new writing will not be of such a nature that it can be understood. The slate endures; but the thought and the writing are always new. Yet, when such writing is of an *impressive* nature, it is like that of a palimpsest; though apparently obliterated, its lines remain, and their meaning can be recalled to memory as often as the occasion may demand it.

The "muddle" of which I have spoken is oftentimes so great that no sound of any kind can ensue, the rhythmic flow of sound-producing streams having been disturbed and prevented from assuming the necessary shape for their formation into proper sound-waves by this hasty mode of thinking. The consequence is a hiatus in the natural flow of speech, which prevents the thought from materializing in the shape of the word intended to be spoken. This hiatus the victim of such precipitate mode of thinking generally attempts to bridge over by spasmodic efforts, which but serve to aggravate the situation, increasing, as they do, the disorder in the sound-producing lines.

Stuttering being caused by a disorder in these lines, the remedy is to again restore them to order. The disorder having been caused by a too hasty mode of thinking, superinduced, as a rule, by a desire *not* to stutter, or a *fear* of stuttering, the remedy lies in allaying this fear. The fear of stuttering, or the anxiety not to stutter, which obtains while the speaker is producing thought, *itself being thought*, and coming on top of the thought intended to be uttered, brings about, or at least aggravates, the very difficulty he was trying to overcome. Mere thought may wander off and again return to its theme, unrestrained, and without causing disturbance; but thought which is to be *vocally* uttered must strictly adhere to its subject. There is no impression to be made by the former which must remain until it is released by vocal sound; impression and expression being almost simultaneous. In place of making a spasmodic effort, therefore, the stutterer should endeavor to be calm, and to then calmly *think* the word or sentence over again which has become a stumbling-block in his way. After doing so, he will have no trouble uttering it.

The fact that stutterers experience no difficulty in singing is a proof of the correctness of these assertions. While singing, the performer's streams of life and organs of speech are all *tuned* to one harmonious measure. His frame of mind being securely in accord with his theme, his thought, devoid of fear, flows evenly along with his song. There is no occasion for haste or trepidation in this instance,—there cannot be, haste being the opposite to and the enemy of harmony, the latter meaning a continuous return of the same measure and the same mode of breathing, the former irregularity and disorder in the mode of breathing.

Besides, song, belonging to the pharynx, is spiritual; it is of our inner nature, and therefore restful and continuous. While speech, which belongs to the oral cavity, is material; it is of our outer nature, and therefore subject to every impression, influence, and consequent change. Elocution, declamation, or recitation, on the other hand, partake of both our inner and our outer nature. They belong in part to the pharynx and in part to the oral cavity.

Experiments may be made by means of making these respective parts rigid which will establish the correctness of these assertions.

These experiments can also be made by the application of mechanical pressure. When pressing your hand or fingers against your throat you will be unable to speak, though it will not prevent you from singing. By pressing them against the back of your neck you will be unable to sing, though you may speak. By pressing them against either side of your neck you will be unable to recite, though you may both speak and sing. The slightest pressure, even, will produce these results. Let me remark, however, that unless the *thought* of the performance accompanies it, a mere mechanical pressure will not suffice.

That *thought*, improperly exercised, is the cause of stuttering or stammering, obtains from the fact, that the utterance of the singer, elocutionist or actor, being a matter of memory, and not of original thought, is *not* subject to these troubles; though the utterance of the same persons while speaking, and in so doing, *thinking*, may be subject thereto.

Not appreciating its significance, I used to laugh with everybody else at the anecdote of a stuttering boy in an apothecary shop, who had been sent down after some article in the cellar. Returning, pale, trembling, and *stammering*, his master cried out, "Sing, sing!" whereupon he delivered himself thus:

"Der spiritus im keller brennt,Und alles steht in flammen."("The spirits, master, are aflame,And all things are a-burning.")

In a recent number of *Cosmopolis*, Prof. Max Müller said:

"Charles Kingsley was a great martyr to stammering, and it was torture to him to keep conversation waiting until he could put his thoughts into words. Singularly enough, at church, Kingsley did not stammer at all in reading or speaking; but on his way home from church he would say to one with whom he was walking: 'Oh, let me stammer now; you won't mind it!'"

While his thoughts were concentrated on his subject, which had probably been elaborated beforehand and was expressed in rhythmic language, besides being obliged to speak slowly and deliberately so as to be heard and understood, he experienced no difficulty. Still, he was under a restraint. As soon as he was by himself again, he commenced to think impulsively, as probably was his habit, and gave vent to a torrent of thoughts, which overleaped each other like waters rushing through a broken dam.

There are two main forms in which this trouble manifests itself. The one is a surfeit, a crowding together of sounds, all of which want to come to the surface at one and the same time, like a crowd of people during a panic trying to rush out through the same door, thus causing a jam. This form, creating a hiatus in vocal utterance, is generally designated by the term "stammering." That which is called "stuttering," on the other hand, consisting, as it does, in a repetition of the same sound, is due to the opposite cause. While the former is due to too great an effort, this is due to a paucity of effort. The sound-furnishing element is not under control; it leaks out against the will, it runs away with you. Hence a repetition of the form once assumed, in consequence of a lack of nerve force, of a rein to keep it in check, of a brake preventing it from rushing downhill with you; in contradistinction to the act of stammering, in which the brake had been too forcibly applied, the watch wound up too firmly and beyond its requirements.

In the case of stammering the impression has been too quick in shaping itself into words; in the other it has been too slow in so doing. In the former case too many moulds have been formed for proper impression; while in the latter the sound is spoken before the mould has been properly and *completely* formed; that part only which had been formed being uttered and repeated. In the case of stammering there is a surfeit of impression but a want of sound; in that of stuttering there is a want of impression but a surfeit of sound. A stammerer is one who takes in too much, a stutterer one who takes in too little, air for his hasty way of thinking.

When this trouble happens with one and the same person—as it sometimes does—it first assumes one shape and then the other; it turns a complete somersault in so doing. The balance, the equilibrium, the point of gravitation, previously overleaped on one side, is again overleaped, and the person lands on its extreme other side. While a stammerer he had too much ballast on board, now he has too little.

A stammerer can return to the point of gravitation by throwing some of his surplus ballast overboard. *His tongue being tied to his lower jaw, in which position he is constantly taking in more air than he needs, he must raise it in order to let the surplus out from beneath the same.*

A stutterer, whose tongue is running away with him, owing to an insufficiency of ballast, must take in enough (inspire sufficiently) to bring him back to his point of gravitation. *His tongue is in a loose state of elevation, in which position the air is constantly streaming out (expiring) from beneath the same.* He must *lower* it to have *his* balance restored, as in so doing the air will stream in over and above the tongue until the equilibrium has been restored. In other words, the person who is thus agitated must calm himself, he must relax from an overstrain in either one direction or the other. The diaphragm, holding the balance of power, will be found to be in as uncontrollable a condition as the tongue, *with which it always acts in unison.* In restoring the tongue to a normal condition we restore the diaphragm to a normal condition.

The institutions for the cure of stuttering, stammering, and intermediate stages of the same trouble, attempt to bring about a state of restoration of the disturbed balance by means arrived at through experience. The real cause being unknown, the remedies must necessarily be restricted. If persons thus afflicted will take their own cases in hand and treat them in conformity with the precepts here laid down, the chances are in favor of their being cured where no other remedy had been of any avail.

As the preceding remarks have been made from the point of view of an English-speaking person, the standpoint of a German being diametrically opposite, the same must all be reversed to fit the case of a German, in so far as locality is concerned. *For stammering, the tongue of a German is closely wedged in, in the direction of the roof of the mouth; for stuttering, it is loosely pointing downward.* This is owing to the fact that a German inspires from under and beneath, and expires from over and above, his tongue; just the reverse of the manner in which this is done by an English-speaking person.

In order to efficiently cure the trouble of stuttering, it is necessary that the act of breathing and sound-production should be closely studied with every separate nationality, as these processes differ with all nationalities; this difference being very pronounced as between Germans and Anglo-Saxons. For an American to go to Germany, therefore, to be cured of this trouble, is as false a step as for a German to go to the United States or England for this purpose.

While I have in the preceding endeavored to give an account of the general causes which result in stuttering, I have not touched upon such special causes as are directly connected with the character and origin of vocal sounds; the explanation of which must be postponed to a future period.

THE CATHODE OF A VOCAL SOUND

By an accident, in some respects not unlike the one which drew Roentgen's attention to the light by whose aid we have learned to look into and through opaque bodies, I (myself an accident, an appearance on and soon to be a disappearance from the illuminated surface of the earth) have discovered eternal laws, by whose aid we shall be able to comprehend much of what has heretofore been as a closed book to us, regarding our physical and psychical nature and the exercise of our faculties and functions.

During my endeavors to overcome the difficulties which my German tongue offered to the perfect pronunciation of the English "r" sound, and during an almost frantic effort on one occasion at so doing, I was amazed by the fact that while one "r" came to the surface from over and above the tongue, another made its appearance from under and beneath the same. The latter was the "r" of the voice of the œsophagus. Of all this, however, I have spoken at length in my previous publication.

Though it occurred to me at once like a flash that this was a revelation of the greatest importance, its real significance was only made clear to me in the course of time. No matter how I view it, as time progresses it assumes greater and greater proportions. There is no event in the history of man which appears to me to be of greater significance. Through this "accident" I was induced to look closer and closer into my inner nature, where, to my amazement, I found that a world, apparently silent and mysterious, and supposed to be unapproachable, was the abode of numberless physical and psychical phenomena, clearly defined and definable.

The "r" which came to the surface from beneath my tongue by way of the œsophagus was the cathode, the negative end of this sound. The *product* of its combination with the *simple* "r" (which came to the surface from over and above the tongue by way of the trachea) I had hitherto produced when attempting to speak English, was the *vocal* "r" sound of the English language; the "r" I hadhitherto produced having been the anode—the positive and first part of this sound only. As Roentgen's cathodic light has illuminated the physical body, so have cathodic sounds illumined for me the spiritual body of my mundane existence. I am endeavoring to show my fellowmen this "new light," whose lustre, also invisible on ordinary occasions, when once seen is so great that it will never again fade from the memory of the beholder. As time progresses, it will continue to penetrate ever more deeply into regions hitherto considered to be impervious to any kind of light; regions whose phenomena have been called supernatural, or, at least, beyond the sphere of the knowledge of man. All other anodes or cathodes of which we have obtained any knowledge belong to physical phenomena only. The cathode I have discovered belongs to our spiritual life, being a part of a living vocal sound.

Think of it! To be able to divide the essence of life and to obtain two *living* parts, each endowed with a life of its own! This is a nearer approach to the knowledge of life than any ever attained before. A *vocal* sound is an entity. From entities we cannot learn anything. They are phenomena complete in themselves. Regarding their innermost nature, they have always been to us as a closed book. They offer us no vantage-ground; no opening, no breach, through which we can enter into the mysterious process of their existence. No matter whether such life or existence be that of the minutest parasite of a minute vegetable growth, that growth itself, or the giant of the forest; whether it be that of a microbe or the microbe of a microbe; whether it be the essence of a thought, a sigh, a tear, a look, a vocal sound, or of a human being—their innermost natures are all alike mysterious to us. I have succeeded in analyzing a vocal sound, and this apparently simple proceeding has opened up to me endless vistas in endless directions. I have reduced this entity into its natural elements, and have again put these together. After resolving it into two lives I have again formed it into one. I can bring about this analysis as well as this synthesis at will at any time.

All know what is meant by vocal sounds, yet few, I repeat, know what are simple sounds, though constantly used by everybody while whispering or uttering exclamations, while surprised, alarmed, frightened, etc. My accomplishment, therefore, is but the *recognition* of the nature of a thing constantly before us and brought to our consciousness through our ear.

Simple sounds are the anodes, the beginnings of sounds. There is no life in them, no rhythm, no melody, no light, no grace, no beauty. These are imparted to them by the fusion of the cathode element of vocal sounds with this, the anode; the spiritual with the material. The anode is formed first. It is the passive element, the female, the patient, the waiting, which must have been before the male, the impatient, the aggressive. The thing to be fructified must have been before that which fructifies.

The anode is quiescent until the cathode comes along, joins it, and infuses life into it. The creation of a vocal sound is an act of generation. The cathode, after overwhelming the anode, penetrates it and diffuses itself throughout it, and thus forms a union whose result is the production of a vocal sound. Similar unions between anodes and cathodes are formed a myriad-fold every moment during time's progress, and result in the creation of an electric spark, or a succession of sparks, called an electric light, or any other light or fire, or of a thought, or of the embryo to a new life of any and every description, etc.; while a discord, a stutter, a *smouldering* fire, the sight of a thing too dimly seen to be recognized, a cut or broken limb, a suspense, a disappointment, a *suppressed* action or passion, etc., are anodes not joined by their cathodes. By the juncture of a cathode with an anode we exercise our faculties, we become conscious of a sight, a sound, an odor, a taste, etc.; the anode being vested in the thing to be seen, heard, smelled, or tasted,—the cathode in ourselves.

While the anode of a vocal sound may be uttered audibly, the cathode, by itself, cannot be uttered—the spiritual cannot be materialized except in conjunction with the material. The anode, the physical, is inert until the cathode, the spiritual, has formed a juncture with it, has been alloyed with it. Every phenomenon of which we become conscious is the result of a process of this nature. The more perfect the union, the more perfect the outcome or result, the phenomenon.

In our ordinary speech this alloy, this union, is of a mutable and evanescent, in oratory and song it is of a more continuous and lasting, nature. With persons speaking a foreign tongue, and with the deaf, it is superficial, imperfect; in many cases, in fact, we hear only anodes, no union having been effected. The amalgamation, the alloy of the finer with the coarser, the higher with the lower, the spiritual with the material, is not at all or but imperfectly performed; the coarser element prevails and makes its presence felt in every utterance. The more perfect the union between anodes and cathodes in vocal utterance, the higher will be the performance, the more perfect the speech, the more beautiful the song, the more stirring, the more soulful; the nearer they come to our hearts.

How do I know all this? I will tell you: By watching the *beginning* of a vocal sound; the performance actually going on within us, while such sound is first being created. This performance is of an inverse order as between German and English, in so far as the anode for German vocal sounds is located to the right, the cathode to the left. The cathode approaches the anode from left to right; while in the creation of an English vocal sound the anode is to the left, the cathode to the right, and the latter approaches the former from right to left. The location where the union *appears* to take place is in the chest, near the heart; for German sounds, to the right thereof, for English to the left. As a matter of fact, however, it is in the heart itself.

What does the motion in which anode and cathode approach each other—which is not direct as it at first appears to the observer, but vastly circuitous—signify?

The circulation through the vascular system of the elements (of the æther) creating vocal sounds, or the *circulation of vocal sounds*. The proofs that this important fact actually obtains will be furnished very positively and very circumstantially at a later date in connection with that part of these expositions which treats on vocal sounds.

OUR MOTHER TONGUE

Nature will have its right always. What is this right in regard to vocal utterance? It is the manner in which we breathe. When we violate nature's right in our mode of breathing for vocal expression, our penalty is that such expression will not be what it is intended to be, what it should be; the idiomatic expression of every language being the outcome of a special mode of breathing for the same.

All my observations in the first instance owe their origin to the fact that I was breathing in a manner directly opposite to the one in which it was necessary for me to breathe to correctly produce the idiomatic expression of the English language. It was not until after this fact had become clear to my mind that I began to extract from my organs of speech those sounds which appear so abnormally different and "strange" to the ear of the bewildered foreigner, who finds himself completely at a loss how to produce them. The better he becomes acquainted with the language, the more thoroughly he becomes convinced of the fact that his mode of speaking English is different from that of the native-born. Nor will a German *ever* succeed in speaking English as it should be spoken until he succeeds in *reversing* his mode of breathing. He must go straight to the antipodes in sound production; he must stand on his head, so to say, instead of on his feet. I shall fully explain what this means later on.

I venture to make the assertion that no other person besides myself has ever learned to pronounce a foreign language *idiomatically correct,* as I have, by means of applying to his mode of speaking rules based on actual knowledge or scientific principles. In this manner I have succeeded in learning to speak English with less of the tinge of a foreign accent adhering to my speech than usually is the case with foreigners who have commenced to speak it as late in life as I did. I do not say this vauntingly, for I do not consider this accomplishment in itself as of a very high order; but I say it to vindicate my claim that I have discovered the principles on which the production of language is based, and offer my personal pronunciation of the English language to which these principles have been applied as a proof that I have done so. I am still learning, however, for it takes time and practice and a great deal of patience to dislodge the old habit from its wonted haunts and to assign its quarters to a foreign guest. My old familiar dwelling has thus become a lodging for the English language, though I can return to it at will with my old and dearly beloved mother tongue and be comfortable therein.

The foreign guest, however, who came to dwell therein, does not use my native home, in his mode of entering it or going forth from it, in the old familiar way, nor does he use the same apartments for the same purposes. He enters at the back gate while I used to enter at the front; he leaves it at the front gate while I left at the back. He opens his shutters to the east, while I used to look out from the west, etc. Such differences as these in our mode of breathing exist throughout the entire length and breadth of both languages. The sounds we have imbibed in our early youth, however, will always be more familiar and nearer to us and dearer than those of any other language, no matter how closely the latter may enter into our lives and our being at a later period.

NATIONAL TRAITS OF CHARACTER

What constitutes a given number of people a nation, besides their history, their political organization, and the geographical position of their territory? What makes every member belonging to a nation, whether he lives within its territory or has emigrated therefrom, a different being from every member of any other nation? What makes each member of a nation resemble every other member thereof, not only in regard to vocal expression but also in regard to general cast of features, build of body, movements, gesticulations, etc., and in what may be summed up as national traits of character?

No one will deny the fact that such differences exist, as between Germans, Frenchmen, and Englishmen, for instance. This difference is not racial, as they all belong to the Caucasian race. It can scarcely be climatic with nations whose territory is adjacent to each other; nor is it likely to be religious, historical, or political. There is nothing very decidedly different in the situation and composition of these various nations and the individuals of which they are composed, except their *language*.

I maintain that language is not only the main point of difference, but that it is the cause and origin of all other main points of difference. As language is the main gift which distinguishes men from animals, so it is also the principal distinguishing mark as between one nation and another. I maintain, and expect to prove, that the language—that is, any specific language—acquired in childhood becomes an integral part of a person's organization, as positively so as any of his other natural faculties; and that he cannot change it, that is, *in an idiomatically correct manner*, without changing, to some extent, the drift of his entire organism. As soon as I began to succeed in speaking the English language as it is spoken in this country, idiomatically correct, I changed my nature, to some extent, from that of a German to that of an American; nor is it possible to learn to speak any language idiomatically correct without undergoing a similar change. Not alone my mode of vocal expression, but my motions, my habits, nay, my very *features*, yes, even my way of *thinking*, in some respects, have been subjected to such a change; modified, of course, by heredity, previous habits, and the constant reversion of all this by the frequent recurrence to my native tongue. In using the term "idiomatically correct" I mean of course that mode of expression which is peculiar to a language, its general cast, and which is representative of its genius and spirit.

To what do I attribute so powerful an influence?

It is not easy to say this comprehensively in a few words. I will say this much, however: That, language being the outcome of streams of the vital fluid passing into and out of our composition in a systematic manner, each system varying with every other system, our vital organs are differently affected, in conformity with the manner and the rotation in which these streams reach these different organs; in other words, in conformity with the manner in which we breathe for our language. This influence is not confined to the vocal expression of a *nation*. It is influential with and extends to the special mode of vocal expression in separate districts, provinces, localities, and cities; nay, it extends to families and single members belonging to such families, each separate member's expression being the product of his special mode of breathing, and differing in some respects from that of every other member of the same family; *such difference in the mode of breathing being the reflection of every individual soul.*

The bent of the soul in *individual* cases determines the flow of these streams, the same as the bent of the *national* soul determines the same for the entire nation. Or, which perhaps would be more correct, the flow of these streams determines the bent of the individual as well as national soul. The influence being reciprocal, it would be difficult to state, as it is with all matters of this kind, *which* preponderates, *which* gives the first impulse. It is of the same order as the old question (never to be solved) aptly expressed in the homely query, "Which was created first, the hen or the egg?"

It is interesting to note the manner in which the vital streams affecting the character of the two peoples in regard to whom I have had the opportunity for many years of making my observations, the Anglo-Saxon and the German, take their course. With the former the point of gravitation is located in the abdomen; with the latter in the thorax.

This gives the Anglo-Saxon a circuitous route for his expression in coming to the surface; his mode of respiration being the following:

He inspires into the thorax posteriorly, next into the abdomen anteriorly. He then expires from the abdomen posteriorly, and from the thorax anteriorly; vocal expression accompanying the last movement.

A German's mode of respiration is as follows: He inspires into the abdomen posteriorly, expiring from the abdomen anteriorly; he then inspires into the thorax anteriorly and expires from the same posteriorly, the latter movement only being accompanied by sound. You will notice that in the former case the breath to be expired and to be accompanied by sound has been held in the thorax until the abdomen has gone through an inspiration and an expiration; while with Germans, inspiration into the abdomen as well as into the thorax are succeeded by expiration from the same, a direct proceeding as against the indirect of the Anglo-Saxon. Thus the former secures a force reserved and held and to be drawn upon as it is needed, while the latter pours forth his vital force in a continuous stream as soon as it is engendered.

The point of gravitation determines the mode of breathing and the production of vocal utterance. With Anglo-Saxons, the point of gravitation being located in the abdomen, their speech tends from below, upward; with Germans, the point of gravitation being located in the thorax, their speech tends from above, downward. The direction of Anglo-Saxon expression is from the abdomen, where it has its root, to the thorax; that of the German is from the thorax, where it has its root, to the abdomen. It will scarcely be necessary for me to say to the reader, over and over again, "Try this," "Try that"; I wish it to be understood, once for all, that this recommendation is to be tacitly implied as accompanying every statement, every proposition, every assertion I make. Personally I can go through any one and all of the performances at any time and at a moment's notice. In making these experiments, speak or sing *after* breathing in the prescribed manner. The prescribed manner being the one in which the *impression* is made and from which the *expression* is produced as a matter of course and of necessity. An Anglo-Saxon will not be able to utter a word spoken or sung in *his* language after breathing in the *German* fashion, nor will a German be able to do so in *his* language after breathing in the *Anglo-Saxon* manner. Change either manner of breathing but in the least, and you will not be able to express yourself in either German or English; but you may thus be able to express yourself in some other language. It is, of course, understood that we breathe into the abdomen through the œsophagus, into the thorax through the trachea.

In trying propositions like the one now under consideration, it may not be easy for persons who have not previously given any thought to matters of this kind to successfully try them. You must give yourself up to these things, must be *at home* for them only, for a period at least, until you have become thoroughly engrossed with them. It is not a study to be superficially attained. You must enter into it with your whole soul, your entire being. If you do, you will eventually become as familiar with the principles underlying these matters as you are with the letters of the alphabet, or the figures representing the numerals, and be able to apply the same in as easy a manner and for as various purposes as you do these.

Their *indirect* mode of breathing of Anglo-Saxons produces a deliberate mode of speech; while German breathing, being *direct*, produces a speech as rapid in its formation as in its utterance. *Action being the counterpoise of speech, is of the inverse order of the latter. English speech being slow and deliberate, English action is rapid and direct; German speech being rapid and direct, German action is slow and deliberate.* English character, the same as English speech, is distinguished by patience and forbearance; these, when finally exhausted, are succeeded by sudden and violent outbreaks. German character, the same as German speech, is alternately exuberant and depressed; contented, but also of a disposition to find fault whenever the occasion may arise.

Anglo-Saxons, in consequence of their *indirect* mode of expression, are in possession of a reserve force always at their command, but only called upon on special occasions; hence long-continued forbearance, and then—a blow for liberty.

With Germans, in consequence of their *direct* mode of expression, their vital force is continuously being engendered, and as continuously being exhausted. Hence, they are in the habit of constantly protesting, and as constantly submitting to the *status quo*.

The character of Anglo-Saxons, in viewing things from a practical standpoint, is as far removed from the ideal as it is from the pessimistic. It is neither exuberant, overstrained, exalted, nor despondent; but cool, well balanced, and matter-of-fact. It is not like the German:

"Himmelhoch jauchzen, zu Tode betruebt."("Raised to the sky with delight;Depressed to the ground with despair.")

A German is influenced according to whether he can or cannot, while losing sight of the real, satisfy his craving for the ideal, for which, in his direct and impulsive nature, he is constantly yearning; which the Anglo-Saxon, seeing it is beyond his reach, abandons as impracticable.

To comprehend the ideal of whatsoever nature, the German, with endless patience, tries to solve the most complicated problems; after solving them he is often satisfied with the result in the abstract; while the practical Anglo-Saxon uses this result for his utilitarian purposes. The philosophical German patiently unravels a Gordian knot; the practical Anglo-Saxon, "Alexander-like, cuts it in two with his sword" ("Wie Alexander haut ihn auseinander"). Germans love education for its own sake; it makes of them superior beings, giving them treasures more highly prized than any others, and far more lasting. Anglo-Saxons, on the other hand, get their education for a purpose, and with a view to their worldly advancement. While with Germans education is "Selbstzweck" (its reward consisting in its possession), with Anglo-Saxons its reward consists in its application. The question so often agitated in this country, whether a university education may or may not be of benefit (that is, in furthering his worldly advancement) to any one not intending to embrace one of the learned professions, would never arise in Germany; practical value and education being things apart, the latter taking first rank always and never being subordinated to the former.

Schiller says:

"[Der Edle] *legt* das Hohe in das Leben,Doch er sucht es nicht darin."

("[Our aim should be] the noble to inculcate into life,And not to search for it therein.")

I am inclined to think that the opposite of this is the usual tendency with Anglo-Saxons.

Many other causes might be cited, many other results. These, however, must answer the present purpose, which is, to show that the course taken by the vital streams in breathing, besides affecting their speech, affects the *character* of nations.

All this might be summed up in saying: The point of gravitation with Anglo-Saxons being located in the abdomen, which represents the material side of life, their being is primarily rooted in the material, and reaches the ideal by way of the material. The German, on the other hand, having his point of gravitation in the thorax, which represents the spiritual part of our existence, reaches the material by way of the ideal, in which *his* being is primarily rooted.

I owe the reader an apology for anticipating in using the terms "streams of life" and "the point of gravitation." These are not words without a definite meaning, however; on the contrary, they are of the greatest significance and of a very definite meaning. Still, I must tax his patience for a proper explanation thereof till I shall be able to reach them in due course of time. We cannot approach the steep crest of a hill by a straight line of ascent, but must patiently wind around and around its circumference to be able to finally reach its summit.

THE AMERICAN NATION

It will require but a single example, familiar to all, to still more forcibly show that it is *language* through whose agency national traits of character and physical development are produced. How do you suppose that the wonder has been wrought, and is still daily being worked, of the great mass of humanity reaching these shores from foreign lands being merged into one homogeneous nation? The remark is often made that "it is the climate." If it were the climate, or other conditions specifically belonging to this country, how is it that foreigners coming here at maturity always remain foreigners, while their offspring born and bred here become Americans? Even children born elsewhere, but coming here at an early age, soon become "Americanized," while their parents remain foreigners always. These children must have taken a potent draught, not partaken of by their parents, to not only change their mode of vocal but also of physical expression; nay, the vital expression of their entire being. That draught is the English language. Most foreigners respectively married to an American wife or husband, and rearing a family of American children, remain foreigners to the end of their lives.

It often happens that parents of foreign birth cannot comprehend the character and actions of their own children, who are *so* different, being superficial and frivolous, where they are deep and sound; cool and calculating where they are fire and flame. Yet these children possess sterling qualities of another kind which their parents do not possess.

I call to mind two brothers, sons of German parents, born in this country. With the eldest-born the German influence was potent. He was made to speak German at home and at school, and is to-day, though married to an American, more German in his manner and appearance than American, while his mode of speaking the English language also has something "German" in it. His brother, on the other hand, more particularly reared under native influences, is a thorough American. There was nothing in this case but the influence of language which could have caused this difference. Similar examples might be cited endlessly.

If language is capable of exercising so powerful an influence it must be more than a superficial acquirement. It must be woven into and interwoven with our innermost nature. What is there in the English language to make a German's broad and massive forehead, high cheek-bones, full lips, short chin, and round face, in his offspring sink into narrow forms and long, oval lines? What makes the lower jaw, which in him was short and round, in these children sink down and extend outward, while the upper jaw recedes back? What is it that makes the jovial and happy expression of the German in his children change into features of an impassive nature, from which they are only roused when in action?—features of which it has been said that it is sometimes difficult to know whether they, sphinx-like, cover a happy or unhappy disposition; a disposition sometimes so self-possessed and reserved that its owner might almost reply as Alva did, when asked why he never smiled: "I would not so demean myself before myself as to smile." Yet when such a face (especially when it is a girl's) *does* smile, its passive features are lighted up in a manner so enchanting that its beauty amply compensates for its previous apathy.

I do not wish to say, however, that Anglo-Saxons do not *feel* either joy or sorrow as keenly as Germans do (though I have my doubts even on this score); but they do not carry their feelings with them on the surface. They sink them into that reserve, at once proud and self-possessed, which does not wish others to take cognizance of their private affairs. The nature of the Anglo-Saxon is one of *reserve*, that of the German one of *abandon* and *laisser-aller*. This is not due to heredity in the first instance, but to the influence of language, by which character and habits are formed.

Dr. Holmes relates that, after a protracted search for his son, who had been wounded in the battle of Gettysburg, when at last finding the "Captain" in a transport train, he went up to him, simply saying, "How are you, Bob?" and he replying, "How are you, Dad?"—stating at the same time, "Such is the force of our national habit that, especially in the presence of strangers, we suppress the impulse of our most ardent feelings," or words to that effect. A similar proceeding under such circumstances would be considered "unnatural" among Germans.

Regarding the change of features, as between foreign-born (German) parents and their English-speaking offspring, by which the latter's assume a shape which makes the œsophagus predominate over the trachea, it will be as impossible for these children to speak *idiomatically correct* German as it is for their parents, with whom the trachea predominates over the œsophagus, to speak idiomatically correct English. When my features assume the proper shape for English speech, I cannot produce a single correct German sound, and when they assume the proper shape for German speech, it is as impossible for me to produce a correct English sound.

I expect that this statement will be hotly disputed. The measure of our ordinary mode of listening, however, must not be applied to these matters. In some rare instances the difference is so slight that it takes a very acute ear to notice it.

CENTRIPETAL AND CENTRIFUGAL

While speaking our native tongue our muscles move, our sinews tend, our vessels lean, *our* blood throbs, and our nerves tingle with the essence of our language in *its* direction, and not in the direction of any other language. We not only speak and sing our language, but we gesticulate it, we walk it, dance it, write it, think it, smile it, and sorrow in it. Everything we do is done differently from the same thing done by a person speaking another language. The movements of the muscles of a German are centripetal, while those of an Anglo-Saxon are centrifugal. With a German they close in around the mouth; with an Anglo-Saxon they depart from the mouth upward and downward. Hence the broad features of the German *versus* the elongated ones of the Anglo-Saxon. Look at the old people. The centrifugal action with an Anglo-Saxon even in old age still leaves his form erect, his face serene, scarcely showing a wrinkle, either on his forehead, his cheeks, or around the eyes and mouth. Apart from his bleached hair, he frequently retains a quite youthful appearance. The centripetal action with a German in old age, on the other hand, has a tendency to bend his form and draw it together, and to shrivel up his skin into innumerable wrinkles, so that his mouth often resembles the mouth of a purse drawn close together. This youthful appearance with aged English-speaking people reflects on their customs and their costume, which latter retains much of the tidiness of their younger days. Germans, on the other hand, age soon. This fact is so apparent that they conform their habits and general appearance to their age. They feel old, and unhesitatingly submit to their aged condition. They often appear old when still comparatively young. English-speaking old people, on the other hand, are never too old not to wish to appear young. For the terms "Greis" and "Greisin," which imply a weakened and somewhat helpless condition, there is no corresponding expression in the English language.

Observe a gang of laborers carrying a heavy log. If there are Germans among them, their heads and shoulders will be bent, as well as their knees, resembling caryatides in Gothic churches. *They carry from below, upward.* Those who speak English, on the other hand, will walk with heads erect, straight shoulders, and stiff knees, resembling the caryatides of the Greek temples. *They carry from above, downward.*

The German mode of expression is produced by contraction, expansion, contraction; the English by expansion, contraction, expansion. For the former, contraction takes place *towards* the diaphragm, first upward and then downward; that is, from the feet upward, and then from the head downward. For the latter, expansion takes place *from* the diaphragm, first upward and then downward; that is, from the diaphragm towards the head, and then from the diaphragm towards the feet.

Artists must study these things if they want to get a proper insight into life, and the action of life, characteristic of different nations. The simple study of anatomy gives them no clue to these matters. Everything we do is done differently from the same thing being done by a person speaking another language. The books on physiology do not make mention of these matters. They treat all nations alike. They tell an Englishman that in closing his mouth the muscles of the upper lip by a direct action are first raised and then lowered, while those of the lower are first lowered and then raised. As a matter of fact, the natural tendency with English-speaking people is towards having their mouths open. In closing the same the lower lip is first raised, then lowered, the upper is first lowered, then raised, and again lowered; whereupon the lower lip is raised. This gives three movements to each lip. The natural tendency with Germans is towards keeping their mouths closed. To *firmly* close the same they must raise the upper lip, lower the lower, lower the upper, and then raise the lower. This gives two movements to each lip. These motions are *indirect* with Anglo-Saxons, with Germans they are *direct*. With Anglo-Saxons the lower jaw is the main instrument; with Germans it is the upper. With Anglo-Saxons the lower moves up to the upper; while with Germans the upper closes down on the lower. That Anglo-Saxons move their lower jaw up to the upper, to them will appear as a matter of course; yet Germans do not do this; with them the lower jaw is first raised to be in position to be met by the upper, the latter being lowered from the atlas by motions made by the entire upper part of the head.

During speech the head of an Anglo-Saxon remains impassive; there is no perceptible movement except in connection with his lower jaw. Hence his stolid immovability in contradistinction with the mobility and vivacity of a German, whose entire head, often accompanied by his entire body, appears to take part in his speech. These motions, though fundamental with these peoples, vary with locality, individual character, temperament, etc. A German if he keeps his cranium entirely still will be unable to produce a sound; while an Anglo-Saxon will be unable to produce a sound if he should move it as Germans do. A German's power of vocal utterance lies in the flexibility of his cranium; an Anglo-Saxon's in that of his lower jaw.

An Anglo-Saxon grinds the teeth of his lower jaw, in anger or in passion, or while masticating food, or under any other circumstances, against those of his upper; a German grinds those of his upper jaw against those of the lower.

All motions in connection with vocal utterance on the part of an Anglo-Saxon are of a decidedly larger compass than those of a German; the latter being confined to the slight motions he is able to make with his head, while the former frequently draws down his lower jaw to a very great extent, far more so than a German would be able to draw down his.

The "life" with the German is in the upper, with Anglo-Saxons it is in the lower jaw; the former representing the thorax, the latter the abdomen. While the thorax, as already mentioned, with Germans is the predominating vehicle for every performance of life, with Anglo-Saxons it is the abdomen.

With Germans the lower jaw is the anvil, the upper the hammer; with Anglo-Saxons the upper is the anvil, the lower the hammer; the action, the life, always being with the hammer.

If you watch an American girl chewing taffy you will find her lower jaw going way down, then out, and up again. This is characteristic of the manner in which Anglo-Saxons breathe and speak. The chewing process, owing to the adhesion of the taffy to the teeth, together with the greater flexibility of a girl's jaws, brings out these features more strikingly than under ordinary circumstances. In chewing taffy the lower jaw (the hammer) meets with some difficulty in making its movements; it is therefore lowered as much as possible, so as to be able to more effectually close in with the upper (the anvil). A German girl's movements under similar conditions are restricted, being largely confined to the upper jaw, which cannot be raised to any great extent.

An Anglo-Saxon speaker or singer makes movements similar to such a chewer of taffy. He draws his lower jaw down and out to make room in the lower cavity of his mouth for the expression of his main sounds. These are the product of the abdominal cavity and find their way out through the œsophagus from *beneath* the lower surface of the tongue. Here they pass the replica and the frænum, which impart to them their rhythmical expression. Any one doubting the correctness of these statements, by making the replica and the frænum, or either of them, rigid, will not, if he is an Anglo-Saxon, be able to produce a single sound; if he is a German, he will still be able to utter his main sounds coming to the surface through the trachea, over and above his tongue. An Anglo-Saxon, on the other hand, may still speak when he makes the vocal cords of the larynx rigid; while a German in that case will be unable to produce any sound whatsoever. To these matters I have already called attention in a previous publication, in connection with the man who was deprived of his larynx by a surgical operation, but not of his power of speech.

A similar experiment may be made in regard to breathing. By making the soft palate, representing the thorax, rigid, you will not be able to inspire, though you may expire. By making the bottom of the mouth close to your teeth (*the soft palate of the lower jaw*), representing the abdomen, rigid, you will not be able to expire, though you may inspire. With a German the precisely opposite facts prevail. By making the soft palate rigid, he will stop expiration; by making the bottom of the mouth close to the teeth rigid, he will stop inspiration.

During vocal utterance, with Germans every superior muscle first moves downward, every inferior upward; while with Anglo-Saxons every superior muscle first moves upward, every inferior downward. This is preparatory and previous to action. *During* action the German opens his mouth, the Anglo-Saxon closes his. Hence the Anglo-Saxon's half-open mouth while in repose, and his almost stern expression while in action, pleasurable action even, which has provoked the witty saying that "Englishmen take to their pleasures sadly."

The abdomen being the centre of gravity for English speech, and the lower jaw being in direct communication with the same by way of the œsophagus, by making the lower jaw rigid you stop the flow of English sounds. The thorax, on the other hand, being the centre of gravity for German speech, and the upper jaw being in direct communication with the same by way of the trachea, in making this jaw rigid you stop the flow of German sounds.

ROTATION OF CENTRIPETAL AND CENTRIFUGAL ACTION

Speaking of centripetal and centrifugal motion as separate actions, there must, of course, be a *rotation* of these actions to produce a *complete* action of any kind. We, however, speak of the one which *prevails* over the other, as *the* action under consideration. Thus when I say a German's mode of eating is centripetal, I say so because the action of his jaws being direct, it is first centrifugal, then centripetal, then centrifugal, then again centripetal. When I say an Anglo-Saxon's mode is centrifugal, I say so because the action of his jaws being indirect, it is first centripetal, then centrifugal, then centripetal, then again centrifugal, and finally once more centripetal. This, with a German, of course, means: Open, close, open, close. With an Anglo-Saxon it means: Close, open, close, open, close. This, however, only gives the main features of an act of eating, etc., as well as uttering sounds; any of these acts, in reality, requiring *eight* movements to carry on one *complete* act. When centrifugal prevails centripetal follows, and when centripetal prevails centrifugal follows. It stands to reason that an action which is composed of open, close, open, close, or close, open, close, open, close, cannot continue in the same rotation indefinitely, but must be complemented by a motion of the opposite nature; such complementary action, however, always being executed inwardly and not outwardly. While the action of the jaws just now described precedes mastication, the inner action complementary thereof is accompanied by the act of swallowing.

Thus with a German there are four movements preceding mastication and four for swallowing; with an Anglo-Saxon there are five movements for the former and three for the latter;

while the act of mastication proper with both nations consists of eight movements which are repeated as often as is necessary for the act of swallowing.

The respective manner in which knives and forks are handled in eating by Germans and Anglo-Saxons, as well as the different manner in which they dance, and the characters they use in writing, might be cited as results of the different modes in which centripetal and centrifugal actions prevail with them. The characters Germans use in writing being centrifugal in their nature and those Anglo-Saxons use centripetal, this can only be accounted for by assuming that the muscular action preparatory to the act of writing in both instances is of the opposite nature.

In consequence of the centrifugal movements of their jaws and lips, the teeth, with English-speaking persons, are always on exhibition; while the centripetal movement prevailing with Germans conceals them. The consequence is that English-speaking people pay the utmost attention to the care and perfection of their teeth, while Germans, in the highest ranks even, frequently neglect them to an almost shameful degree. The direct outcome of this state of affairs is the great advancement which the practice of dentistry has made in this country and in England, while it is one to which, on the continent of Europe, but comparatively little attention is being paid.

With English-speaking people, especially the women, whose lips are more flexible than men's, the teeth of the upper jaw are more frequently exposed than those of the lower, for this reason: The œsophagus being the main instrument for English speech, its sounds, in coming to the surface from beneath the tongue, require the latter to remain in a semi-raised position most of the time; the upper lip, being in the way of these sounds coming to the surface, must be raised for the same reason; in so doing it exposes the upper row of teeth. The lower lip is lowered for the sounds of the trachea for the same reason that the upper is raised for those of the œsophagus. Whenever the upper lip is raised the lower must be immediately lowered, and vice versa. With Anglo-Saxons the main movement is with the upper, with Germans it is with the lower lip. Owing to the centripetal action with Germans, these movements are less pronounced than they are with English-speaking people.

The act of smiling being produced in the same order as that of speaking, the same conditions prevail in relation to the same.

In speaking English you can "feel" that the upper lip is the main vehicle; *it has all the life in it.* In speaking German you can "feel" it is the lower, which for that language possesses the life. If you make the former rigid you cannot speak English; if you make the latter rigid you cannot speak German.

In connection with the movements of the lips it will be noticed that while the upper jaw and the roof of the mouth are dominated by the trachea and the thorax, and the lower jaw and the bottom of the mouth by the œsophagus and the abdomen, the upper lip is dominated by the sounds of the œsophagus, and the lower by those of the trachea. This, however, is owing to mechanical reasons only, as explained, and not to vital causes.

The foreigner who learns to speak the English language ever so well, though he may reside here almost a lifetime, if he does not learn to speak it *idiomatically* correct, will not be influenced by it to any great extent in any of the various manners of which I have made mention, either as regards his features, character, habits, motions, thoughts, etc.; but, in spite of his "English," he will still be a foreigner. This foreigner's children, however, provided he does not influence them to the contrary through pride of his native tongue, and if reared under native influences, will become thorough Americans.

There need be no fear, therefore, that immigration might bring to this country a permanent foreign element. Such elements, when they do come, are of a passing nature. Their offspring, in passing the crucial test of the English tongue, sink the foreigner into the all-absorbing element of the English idiom; and in so doing are merged into and become an integral part of the people of this country. They may come of whatever nation, from whatever land; no matter how they may appear, act, or speak, the English idiom will continue to make them Americans, in their children at least, in the future as it has in the past. There is thus in the centrifugal force which dominates the speech of Anglo-Saxons that which is a safeguard to the homogeneity as well as the institutions of this nation.

An Anglo-Saxon cannot be a bondsman; his language forbids it. The centrifugal force which prevails with him does not permit fetters. The children of all foreigners born here and speaking the English language come under its spell. If language did not have this supreme influence, there is no other influence that would have prevented this country long ago from having become inhabited in special districts with permanent groups of people foreign to its aims and institutions, and alien to its genius, its character, and its customs. In districts where German is spoken as the principal language, as in some parts of Pennsylvania and Wisconsin, it is not, with the native-born at least, the pure German language, but its idiomatic expression is that of the English tongue.

People say, "It is the climate." We have every climate under the sun; yet in all that is essential the man from Maine is as thoroughly American as the one from Texas; the gold-digger in the frozen regions of the Yukon as the man of the orange-groves of Florida or California; the American fisherman on the Banks of Newfoundland as those on the Gulf of Mexico; the man who battles on the plains against the Indians as he who serves under the banner of the Republic and upholds its glory in foreign lands and seas. You can tell an American the moment you look at him. Yet if you ask some of them where their parents were born, you will hear strange tales of lands and peoples across the sea and far away.

Language does not work *every* wonder, of course. The influence of heredity perpetuates that of language; but the latter is the primary influence. Nor can it be denied that *every* foreigner living here for some time, whether he has learned to speak English or not, will, to some extent at least, be influenced by the habits, customs, institutions, climate, and language of this country. This does not detract, however, from the force of my argument regarding language and its influence as the most vital force in shaping a people's characteristic traits, physically as well as spiritually.

There has been of late a great deal of talk and enthusiasm even regarding the desirability of a closer alliance between the two great English-speaking nations; their natural affinity and kinship. This affinity, this belonging together, this being of one family and one stock, is commonly expressed by this term, "English-speaking peoples." That which I have endeavored to explain at length is thus tacitly acknowledged to be correct through the use of this term, which implies that it is *the English tongue* which makes these peoples one in sentiment, in feeling, in their aims and purposes, as it makes them one in their physical appearance, their motions, the exercise of their faculties and functions, etc.

NATIONALITY AND RACE DISTINCTIONS

While the English language makes Americans of all foreigners, it does not, of course, obliterate race distinctions as long as races continue to exist as such. Persons of alien races, nevertheless, when born in this country and reared under native influences, will become "American" in a truer sense than foreigners belonging to the Caucasian race coming here at maturity. I dare say Frederick Douglass was truly more of an American, in all this word implies, than any foreigner who ever came to live here; and so are all the better classes of native-born negroes, in a certain sense, more truly American, this indescribable something which constitutes a nation, than any aliens whosoever.

A gentleman once told me that, travelling on a steamboat on one of the New England rivers, he had been inadvertently listening to a conversation carried on behind him, between what seemed to be two New England farmers. On rising from his seat, he saw that one of the men was a Chinaman, dressed like the other and conversing precisely as he did.

Seeing an acquaintance, he pointed out the Chinaman and asked if he knew who he was.

"That's Jimmy O'Connor; he's from So-and-so."

"I mean the Chinaman."

"Yes, the Chinaman; that's him. You know he was picked up at sea, when still a baby, by a New Bedford whaler, and was brought up in the captain's family, who adopted him. He's as good a farmer and as true an American as you can find anywhere."

These studies are meant to be purely objective, and have no concern with politics or policies, regarding undesirable immigration, or issues of a similar nature. But language is nationality, and nationality language, always, in the first instance; and the purer a language is spoken, the truer, purer, and better such nationality will be expressed and represented by those who thus speak it. What an incentive to aim at the purest and best expression of language, for any people! But it will be said that language is subject to change. If it is, so will the people who speak it to some extent change with it. Such change, however, is in its dress, in words mainly; rarely and at long intervals, and under very peculiar circumstances only, in its expression. As a matter of fact, I doubt whether a change of the *idiomatic expression ever* takes place.

The difference existing between the English spoken in the United States and the mother country might be cited as an example. The idiomatic expression is precisely the same. But the necessary self-reliance of the first settlers, the privation, the barter and exchange, the vast extent of the territory of this country, the greater independence enjoyed by its people, etc., might be named as reasons for the greater dash and freedom, together with a possible want of culture, as compared with the language spoken by educated Englishmen, prevailing in its utterance.

The same influences prevail regarding the general appearance, motions, and characteristic traits of these respective nations. Though closely allied and connected in a specific, and very nearly allied to each other in a general sense, there is that which distinguishes the English of the old world from those of the new, and which can be easily recognized.

Being centrifugal, the English idiom, octopus-like, embraces anything and everything that comes within the radius of its omnivorous capacity, without, however, losing its original character. It is like a fisherman who has hung out his net in the ocean, taking in all that comes along; or like the sea itself, greedy without end. It has no scruples about roots and construction, but construes everything according to its wants and adapts it to its uses as it comes along from any quarter.

These adopted children, these waifs, however, it must not be lost sight of, before they become integral parts of English speech must submit to a change of their original idiomatic expression. No matter who came—Celts, Romans, Angles, Saxons, or French—the people of the British Islands, while adopting their *terms* of expression, remained true to their original *idiomatic* expression. As this country absorbs people from the whole world and makes one homogeneous American nation of them, so has the English language absorbed, and is still absorbing, words from every other people's language, and has transformed them into one homogeneous language of its own.

Comparative philology, if it wants to accomplish that which would be most worthy of its efforts, will have to come down to these strong and basic roots of language.

The German language, whose idiomatic expression is centripetal, on the other hand, does not possess the same capacity for adopting foreign words and adapting them to its idiom. When it does adopt them, as, for instance, those of French origin, they are pronounced, not in the German, but, as far as the German people are capable of so doing, in the French manner. They could not, in fact, be pronounced in the German manner, the German language being a close corporation, so to say, which does not admit of any foreign shareholders; while the English language is a company open to all comers. While it is the endeavor of Germans to *purify* their language by expelling as far as possible any foreign word and element therefrom, Anglo-Saxons are constantly adopting new words from foreign languages. It would be equal to the labor of Sisyphus for Anglo-Saxons to endeavor to purify their language from foreign words, in the same sense that Germans are attempting to purify theirs.

It appears to me that the capacity of England for successful colonization is largely due to the centrifugal force inherent in its language, while the want of success of Germany for the same purpose is due to the absence of this force. Anglo-Saxon government tends toward decentralization, German toward centralization. I say this in spite of the fact that Germany is still divided into many principalities; the fact of its adherence to this undesirable condition being a proof of the correctness of this assertion rather than otherwise—Germans not being able to readily get out of that in which they are once rooted. In regard to governing peoples in distant territories or colonies, this tendency is of importance. English government, being undemonstrative, is more effective than German, which is demonstrative, meddlesome, and therefore offensive; the former being material and practical, the latter immaterial and inclined to be visionary.

In a word, where are we to find explanations regarding national traits of character except through inner motive powers, productive of results individual as well as national? There is no factor which exercises an influence upon a nation as a unit so wide in extent and of so powerful a nature as that of language. It is the *only* motive power, in fact, which every member of a nation shares with every other member thereof, but not with any member of any foreign nation.

IDIOMATIC EXPRESSION

Although it is a well known fact that every language has an idiomatic expression, an intonation of its own, I am not aware of any attempt ever having been made at definitely stating what such expression, or intonation, really consists in; and in what respect it differs, as between one language and another. Yet this fact should be the most important of all in connection with ethnological studies. It is necessary to know what a people's idiomatic expression is before we can begin to make a study of its language, in comparison with that of any other people, by which we may expect to arrive at conclusions of any real value in an ethnological sense.

In comparison with idiomatic expression, the study of the roots of words and their derivation, it appears to me, is of but secondary importance; idiomatic expression being the *kernel* in which the tree of national expression had its incipiency, its origin. It is the life which pulsates through its veins, in which it has its stay and maintenance; the nerves which tingle with its intelligence, its genius, its soul. Take away this soul, and it ceases to exist. For every language there must have been a strong impulse making an impression before there could have been any expression at all. This impulse must have been of so powerful and continuous a nature as to have left its impression upon the minds of a sufficiently large number of people to form the nucleus for the expression of a specific language, and, in so doing, constituting such people a nation.

I have already stated that it is *motion* in the first instance which superinduces a specific mode of breathing and consequent expression. It is to motion, then, that we must ascribe the first impulse. Such motion may have been active as to defense against enemies, wild beasts, or the elements; or it may have been passive, consisting of the continuous noise produced by the motion of the sea, tempests, or thunder-storms, making a great and lasting impression. Then, again, the influence may have been of a peaceful, balmy, beneficial nature, as with people living in security, in a mild climate and on fertile lands. The stronger the expression of these movements, the stronger the impression they made and the more powerful the expression of the language; the softer and more harmonious their expression, the softer and the more rhythmical the expression of the language. These influences made their first impression by superinducing a mode of breathing in conformity therewith.

Thus sounds giving expression to pain, perhaps, in the first instance, or to sorrow, joy, surprise, etc., were made in conformity with this, their specific mode of breathing. These outcries, consisting of syllables, grew into words and sentences, which, being uttered in conformity and sympathy with their special mode of breathing, created a specific idiomatic expression. The same process, from its first inauguration, and with but slight alterations, has been practised and persisted in by the same people from the beginning to the present time. With the English people, as already mentioned, no migration, no invasion, no conqueror, no matter how powerful, has been able to swerve it from its path. The *most* these invaders could do was to graft some of the expressions in which *their* ideas were clad, some words, on to this aboriginal stem. This stem was so strong in its primeval conception that it could bear all these exotic graftings without losing its character, absorbing all, welcoming all beneath the widespread roof and homestead of its branches. It proved its superiority over the idiomatic expression of these foreign tongues by its survival, as the fittest.

[Before proceeding further, I want to remark: these studies having been made from an Anglo-Saxon point of view, it is just possible that a preponderance of observations may have been made on that side; while, if they had been made from a German standpoint, the preponderance most likely would be on that side. This, no doubt, will be the case should I at any future period be able to write all this, as I intend to, in the German language.]

What is this original sap in the English, and what is it in the German language?

The aborigines of the British Isles, living apart from their continental brethren, became possessed of an idiom different and apart from any other. It was the idiom of the *sea*, by which they were surrounded; the motion and commotion of the waves, the surf, the incoming and outgoing tides, their undertow and overflow; the waves advancing toward the shore, their breaking against it, and their final retreat from the same.

The English language is a raft living upon the ocean. You can *hear* the waters rushing through it and on to the shore and back again. You can feel the waves rising up to gigantic heights, and then falling to and below the level of the sea. You can feel the undertow in its reserve force, quiet and subdued like the lull before the storm, yet capable of almost any demonstration. You can feel all this in the strength and vigor of its diction as expressed in its prose and poetry. This is not a mere poetical conception, but a truth capable of actual, practical demonstration.

While reading poetry or prose, or while singing, fancy seeing in your mind's eye the ocean with its waters in commotion, either the open sea or the surf near the shore, and you will *feel every word you utter mingle with its waves. These pictures will never disturb your fancy, but will associate with it in perfect harmony.* Now substitute for the picture of the ocean and its tumult some rural picture, as of a field of grain or the branches of trees tossed by the wind, or the flow of a river, or even that of the sea itself when perfectly calm. Keep such picture before you exactly as you did that of the sea in commotion. While reading, speaking, or singing English you will not be able to *hold* such picture; *it will soon disturb you, and to such an extent that you must cease thinking of it, or be obliged to stop your reading, singing, etc.*

The impression made by the ocean, in fact, is so great that it dominates the *thought* and the entire being of English-speaking people. This is the case to such an extent that if you continue to persistently *think* of any other image than the ocean, even without uttering any sound whatever, it will so greatly perturb you that you will be unable to continue thinking at all. You may, on the other hand, continue to think for an indefinite period of the image of the ocean without experiencing any disturbance whatever.

While the basic element of the English language is closely affiliated with the ocean, that of the *German language* is affiliated with the *woods, and the blowing of the winds.* In their habitation in the forest, the wind made so deep an impression on the primeval inhabitants of Germany that you can feel its *soughing pervade all German diction.*

If you are a German keep the picture of the woods before you and the soughing of the wind through the tree-tops, and it will harmonize with German thought and diction. Substitute a picture of the ocean for it, or almost any other picture, and you will not be able to vocally utter German thought, nor will you be able to continue thinking in the German language at all.

In place of conjuring up these pictures in your mind's eye you can substitute *real* pictures representing these scenes, and while contemplating them the effect will be the same.

After pursuing the picture of the ocean for a while, say: "English;" after pursuing that of the woods, say: "Deutsch;" either will come quite naturally, but you cannot reverse them. If you attempt it, these words will not be forthcoming.

While with English diction there is *a pause and then an emphasis* as of the waves coming on and then breaking against the shore, so, with German diction, there is an *emphasis and then a pause*, as of the blowing of the wind succeeded by a calm. These, in a word, are the characteristic elements in the idiomatic expressions of these peoples; English idiomatic expression being *low succeeded by loud*; German, *loud succeeded by low*.

The influence of the ocean with its continuous uproar formulated the speech and character of the English nation into one of strength and reality, with its centre of gravity in the abdomen. The peaceful influence of their habitation in the woods, together with the impression made by the wind, the singing of birds, etc., formulated the speech and character of the German nation into one more of ideality, with its centre of gravity in the thorax.

The fondness of the English for the sea, their supremacy thereon, etc., need not be amplified upon:

"Wherever billows foamThe Briton fights at home,His hearth is built of water."

The fondness of the Germans for the woods is equally noted: Der "dunkle," "zauberische," "geheimnissvolle," "heilige"—

Wald (The "darkly deep," "magical," "mysterious," and "sacred" woods) are but common expressions.

There is not a word in the English language of the same significance as that of "Der Wald." It embraces many ideas, of which the words "the woods" and "the forest" are not expressive. These, in a literal translation, find expression in the words "Das Gehoelz" and "Der Forst," which are of a more realistic nature.

The English language, on the other hand, is full of expressions applying to nautical matters and to the sea, for which there are no adequate expressions in the German language.

The fondness of the present Emperor of Germany for the sea must be attributed to the English blood flowing in his veins. While it is his desire to create a powerful navy, the people of Germany are indifferent to, and obstruct rather than assist, the accomplishment of this desire.

Idiomatic expression, the soul of language, has its incipiency in the *soul* of a people, and may pervade it for centuries before the *body* of the language, the *words* in which its thoughts are clad, makes its appearance. It must have taken many centuries more before these words grouped themselves into sentences and assumed the shape of speech. The words may change, but the idiomatic expression will always remain the same.

So, also, must the soul of man have had existence for an indefinite period of time before a body was formulated to clothe it in. The spiritual cell, if I may be permitted to use such an expression, must have existed before the material; or, in other words, the spiritual cell must have made its appearance long before the material cell *commenced* to make its appearance.

RELATIONSHIP SUPPOSED TO EXIST AS BETWEEN THE GERMAN AND ENGLISH NATIONS

It is a common saying that there is a close relationship existing between the German and English nations. There is no greater fallacy than this. I contend that this relationship is of a very distant order, consisting, as it does, merely in words, or, as I have said, garments loosely flung around the sturdy, strong, and unalterable stem of English idiomatic expression. In every other respect there is a great dissimilarity and antagonism even, existing between these two peoples. If there is any analogy existing between them at all, it is one of opposition; one that is based on the idea that extremes meet (*les extrêmes se touchent*), their poles being diametrically opposed to each other.

There is no more relationship existing between (Anglo-Saxon) German and English than there is between (Norman) French and English; the German, French, and English languages each possessing their own especial and unalterable idiomatic expressions. Whatever foreign words either of them adopt must be subjected to their idiom, or keep floating along as best they may in their original character.

The entire aspect of these three nations, the French, English, and German, points to the fact that there must be a radical difference in their vital mode of existence. Just what this vital mode consists in, in respect to the two latter nations, I expect to still further establish in a future publication. Both languages traverse nearly the entire range of the vital organs in opposite directions. Hence the strength and also the weaknesses of these languages, as compared with other languages which, extending from side to side, have a smaller compass but a comparatively purer range of sounds. Regarding other nations and their languages, I trust others, thoroughly familiar with the same, by applying to their investigations similar principles, will establish similar facts.

Owing to its centrifugal tendency, it is necessary for English vocal utterance to open the mouth much wider than it is for German. Let a German open his mouth no farther for the enunciation of English than he is in the habit of opening it while speaking his own language, and he will not be able to utter a single sound. The same result will obtain when an Anglo-Saxon attempts to speak German on the same basis that he is in the habit of speaking his own language. Owing to the centripetal tendency of the German language, the mouth in speaking German is but slightly extended. That this respective widening and narrowing of, not only the mouth but of every other channel employed in bringing about vocal utterance, must tend to exercise a marked influence on Anglo-Saxon and German features will be obvious. The consequence is that the mouth of English-speaking persons in thus being extended has a broad yet narrow appearance, with rather thin and compressed lips, while the mouth of Germans in thus being contracted is comparatively smaller, with full and ripe lips. This feature is in conformity with all other features which, with Anglo-Saxons, are elongated, with Germans contracted.

Experiments regarding centrifugal and centripetal action can be made to good advantage by resting your head sideways on a pillow. In this position during vocal utterance you can *feel* these actions, and, feeling them, "*measure*" them. This mode of proceeding can be successfully adopted in many other experiments connected with these studies. I must warn the reader, however, again and again, that all this has reference only to languages spoken idiomatically correct. It has no reference whatever to foreign languages spoken in the usual mechanical manner.

LANGUAGE AND MOTION

I will now show that motion is the first impulse and primary condition of speech. I will give but a few examples at present, but expect to prove most exhaustively later on that motion *must* precede, or *apparently at least,* accompany vocal sounds *always.*

While standing up, straight, throw out your arms horizontally, then speak English. You will have no difficulty, but you will not be able to speak German so easily. Next, stand as before, and again throw out your arms horizontally, then drop them, letting them hang down close to your body. After doing so you will have no difficulty in speaking German, but you will not be able to speak English so readily. In throwing out your arms in the first instance, your mouth will open, and you will *close* it in speaking English. In letting them drop, in the second instance, your mouth will close, and you will *open* it in speaking German. Now, stand on the tips of your toes, and you will have no difficulty in speaking English, but you will not be able to speak German with ease. Then rest the weight of your body on your heels, and you will have no trouble in speaking German, but you cannot speak English with ease. In standing on the toes the body is extended by centrifugal, in standing on the heels it is contracted by centripetal action. Next, extend your neck, and you will have less trouble in speaking English than in speaking German; then lower your neck, and you will find no trouble in speaking German, but you will in speaking English. These experiments might be amplified manifold, but these must suffice for the present.

The same features of the opening and closing of the mouth in conformity with the position you assume, will obtain in all these instances the same as at first mentioned. It will scarcely be necessary for me to repeat that all this shows that the motion for English speech is centrifugal, for German centripetal. Nor will it be necessary to call attention to the fact that all this tends towards giving Germans a condensed and broad, Anglo-Saxons a lengthy and narrow bodily appearance.

It is, however, a noteworthy fact that with Germans the nearer you approach the sea, the more centrifugal becomes their action and personal appearance. The people of Northern Germany, therefore, though radically differing from them in most other respects, partake more of the general bodily features of Anglo-Saxon nations than those of the South of Germany, who are positively opposed to them.

Upon having ascertained the correctness of these statements by actual experiment, I want to ask the reader how he expects to reconcile these facts with the universally adopted theory that the larynx is the sole instrument productive of vocal utterance. An Anglo-Saxon, when stretching out his arms horizontally, can readily speak English, while a German in the same position cannot utter a sound of *his* language without difficulty. If the larynx in the case of an Anglo-Saxon, under these circumstances, produces vocal utterance, why is it not so easy with a German?

My explanation is this:

By extending your limbs, in stretching out your arms, or standing on your toes, the centrifugal action is instrumental in parting the jaws and giving the tongue an upward tendency. In so doing, the œsophagus and replica obtain ascendancy over the trachea and the larynx. The abdomen (the seat of gravitation for English speech) and its tributaries thus obtain the mastery over the thorax and its tributaries. The former being the main vehicle for English speech, such speech can be produced without molestation. These facts, while favorable to the production of English vocal utterance, obstruct and hinder German vocal utterance.

In lowering the arms or standing on one's heels, thus substituting centripetal for centrifugal action, the jaws close, the tongue assumes a downward tendency. The trachea and the larynx, as well as the thorax (the seat of gravitation for German vocal utterance), obtain the preponderance, and German may be freely spoken, while English is obstructed.

In *raising* the tongue, a free passage to the œsophagus is obtained, while that to the trachea is obstructed. In *lowering* the tongue, a free passage to the trachea is obtained, while that to the œsophagus becomes obstructed. It is necessary, however, to understand that, while English speech is centrifugal and German centripetal, these are *tendencies* only and not permanent *conditions*; centrifugal and centripetal action constantly interchanging and modifying one another. An uninterrupted tendency in one and the same direction, either centripetally or centrifugally, would soon come to an end and produce stagnation, inertia, death. There is no action without a counteraction. Hence, ingoing vocal sounds are counterbalanced by outgoing; the same as ingoing thoughts or thoughts produced by external vision are counterbalanced by outgoing, or thoughts produced by internal vision, etc.

In addition to the parts mentioned, there are many other parts of the body which, subjected to centrifugal or centripetal action, will produce results of the same order as those already mentioned. In stretching out your legs (while in a sitting position), you will find speaking German to be difficult; upon drawing them up, you will have trouble with English. The same results may be obtained, in connection with the toes and fingers, in a number of different ways. From all this, it will be readily seen that all parts of the body are closely related to each other, the tendency of the muscles in one prominent part producing the same tendency in all the rest.

There is one thing which must be mentioned, however. To obtain centrifugal action, it is necessary to *stretch* the part under consideration; the mere extension of a part, without stretching it, will be fruitless of results in either one direction or another; so will the mere contraction of any part be fruitless of results, unless such contraction is complete. You can let your arms hang down alongside of your body and yet speak English easily; and you can hold them out horizontally, and yet speak German easily. In either case the contraction and expansion must be *thorough* to produce results either centripetally or centrifugally.

All persons make similar motions to those mentioned with every sound they utter, though these motions do not appear on the surface; in fact, they could not speak if they did not make them.

I have already mentioned, but want to repeat, that centrifugal action is the cause of the elongated faces, and especially of the elongation of the lower jaw of English-speaking persons. It is also the cause of their semi-parted lips while in repose, showing their teeth, and a full exhibition thereof while speaking; a fact which has caused much merriment to continental nations, and has given rise to an endless number of caricatures of "milord" and "milady" on their travels, etc. It is also the cause of the perfection of dentistry in this country and in England, where the teeth are always more or less on exhibition. In other countries, where they are hidden behind the curtains of the lips, which are usually closed, except while speaking or laughing, this necessity does not arise to nearly the same extent. To the centrifugal force there is also due much of the innate charm and beauty of English-speaking women.

From all this one great lesson may be learned: no matter by what divergent means nature may work its ends, similar results are obtained, though often arrived at by opposite means and from opposite directions. Thus life ever presents to us new forms and features, and ever infuses new interest into what otherwise might become unbearable in its monotony. A better insight into these facts ought to make us feel more lenient towards what appear to us as other people's "idiosyncrasies." It should also have a tendency to prevent us from attempting to enforce to their full extent laws made in conformity with our own desires and inclinations but in direct opposition to those of others (foreigners living among us), whose character and disposition lead them in diametrically opposite directions.

Unless otherwise mentioned, I wish the reader to remember that I am always speaking not only from the standpoint of an American, but *as* an American. The fact of my long residence in this country, where I have spent the best part of my life, in itself would not entitle me to do this, having shown, as I have endeavored to do, that this is not sufficient to change a person from one nationality into another. During my earnest endeavor at fathoming these differences, however, I have been led into assuming the forms which distinguish the Anglo-Saxon from the German. Unless I am with Germans and speak the German language, in my thoughts and otherwise I lead the life of an American.

That my English speech, however (though my friends in their indulgence would lead me to believe otherwise), is not as perfect as it might be, is largely due to the fact of my constantly having recourse to the German language, and that I am thus as constantly led back into these other forms of existence which cannot be indulged in without some detriment and abstraction from either the one or the other. There was a time, in fact, when the transformation I have spoken of was taking place (the disturbance being so great) that I could not speak well either the one language or the other.

I am well convinced, on the other hand, that through perseverance *perfection* in the utterance of both of these languages, for speech as well as for song, and possibly of some other languages besides, may be attained in the course of time; nature being so pliable that, when the required actions are once *fully* understood and complied with, a perfect change may be made instantly in passing from one language on to another. Such changes, in fact, are naturally made by persons who, in their infancy, have been educated in and taught to speak several languages at one and the same time; the material during infancy being so pliable that it can be readily formed into any shape and transformed into any other. All of the preceding also shows that, for every separate idiom, the *entire* instrument must be "tuned" for its production in a given order, and that only when so tuned can such idiom be produced in its entire purity. It also shows that, unless so tuned, the vocal cords of the larynx and replica cease to be instrumental in the production of sound.

An instrument tuned for the production of the English language, consequently, cannot produce German sounds, nor can it produce Romanic, Slavonic, or the sounds of any other language. Sounds, *apparently* the same, of either the singing or speaking voice of various languages are, therefore, *not* the same and are certainly not produced in the same manner. For a German, consequently, or an Italian to attempt to teach an

English-speaking person the art of singing is an anomaly. A foreigner might, with the same show of reason, attempt to teach persons of another nationality the correct pronunciation of their own language. It would be equally false, of course, for an English-speaking person to attempt to teach a German, Italian, etc., the art of singing, unless he had first mastered his pupil's idiomatic expression, or the pupil had mastered that of his teacher.

Many persons are under the erroneous impression that song and speech are performances separate and apart from each other, while they are in reality of precisely the same, though inverse, order. They are of the same order, for instance, as the back and palm of the hand: the former representing speech, the latter song; the external and the internal, or the anterior and the posterior. As the back of the hand, such must and will be its palm; or, as its palm, such must and will be its back.

Conversing with a teacher some time since, she scorned such propositions, saying a person's language had nothing to do with his or her song; the mode of production of the latter being the *same* with ALL nationalities; besides, she had studied the larynx, and knew all about it. This, of course, settled it, and I had not anything further to say.

DIFFERENCE IN THEIR MODE OF BREATHING AS BETWEEN ANGLO-SAXONS AND GERMANS

Anglo-Saxons inspire first into the thorax and then into the abdomen. Germans inspire first into the abdomen and then into the thorax. The former expire first from the abdomen and then from the thorax; the latter expire first from the abdomen and then from the thorax. This, however, gives but a partial account of the process of breathing, and I must postpone a more explicit one to a later period.

To prove the correctness of the above assertion, press your hand against the left side of your thorax anteriorly, and you will find it difficult to inhale. If you press your hand against the right side of your thorax, on the other hand, you will have no difficulty in inhaling. Next, press your hand against the right side of your abdomen, and you will not be able to exhale; but if you press your hand against its left side, you will experience no trouble in exhaling. In pressing your hands one against the left side of the breast and the other against the right side of the abdomen, you will have trouble in breathing.

Pressures produced in the precisely *opposite* manner in every respect, on the part of a German-speaking person, will produce effects of precisely the *same* nature. A German, in pressing the right side of his abdomen, will not be able to inspire freely, but pressing its left side will not hinder him from doing so. Pressing the left side of his thorax will impede his expiration, while the pressing of its right side will not prevent him from doing so. These results will become more obvious when these pressures are continued for some time. All the pressures mentioned are to be applied *anteriorly*. Pressures of the same nature applied *posteriorly* produce opposite results with Anglo-Saxons as well as Germans.

Similar results may be obtained by producing pressures on the median line of either thorax or abdomen, front as well as back. Such will also be the case when pressures are produced on either side from the armpits downward or from the hips upward. More satisfactory results, however, than those obtained through mechanical pressure can be obtained by making the respective parts rigid. It will scarcely be necessary for me to mention all these various causes and consequent results in detail, as any one interested in these matters can work them out for himself from that which I have said.

RISE AND FALL, OR RHYTHM

The thorax is productive of the falling, the abdomen of the rising voice, the former being the representative of the *impression* for sounds, the latter of their expression.

An Anglo-Saxon's voice, inspiring, as he does, into the thorax, and expiring from the abdomen, will first fall and then rise. A German's voice, on the contrary, inspiring, as he does, into the abdomen, and expiring from the thorax, will first rise and then fall.

This is the fundamental cause of the difference between the idiomatic expression of these two peoples, and primarily also of the difference existing between their national traits physically as well as mentally.

Every original word in either of these languages will illustrate these facts:

´ ` ´ ` ´ ` ´ `

Vater, Mutter, Bruder, Schwester.

Take the same words in English, and the accent will be reversed:

` ´ ` ´ ` ´ ` ´

Father, Mother, Brother, Sister

When these and similar words were adopted into the English language, it was done at the expense of their original idiomatic expression. I am speaking of the music, the rise and fall, the rhythm pervading a language, not of time or measure, nor of the intonation, nor of emphasis.

I make four distinctions, and expect to prove that they are the basis of every artistic expression of either speech or song. First, measure or time. Second, the rise and fall of the voice, equal to its rhythm. Third, intonation, which pertains to words in accordance with their meaning. Fourth, emphasis, which has reference to the feelings.

That the human voice is capable of at one and the same time expressing four moods so different from each other, shows that there are various factors (all of a different nature) simultaneously at work producing these different results. To correctly indicate these four characteristics, it would be necessary to mark each syllable in a fourfold manner. I shall confine myself to the rhythm and the metre, and shall mark the former above the line by using the signs for accent (́), and the latter below the line by using those for metre (ˉ ˘).

Right here is the main stumbling-block with persons of either nationality in speaking the language of the other. They will in so doing invariably retain the idiomatic expression of their own vernacular.

The *proper* way to illustrate the rhythm would be as follows:

 ́ ˋ   ́ ˋ   ́
Vater, Mutter, gut.

ˋ ́ ˋ ́   ˋ
Father, Mother, good.

There is always a rise of the voice before its fall in German, and a fall before its rise in English *for each and every syllable*. When a language is well spoken, this complete intonation is always heard. If this needs illustration, which it should not, being so obvious, the poetry of both peoples offers proofs in great abundance. It is a notable fact that, with German verse, the voice for the end syllable always sinks, with English it rises; the former is generally short, the latter long; but even where the word ends with a long syllable in German the voice falls at the end, and where one ends with a short syllable in English the voice rises at the end.

To anxiously count every syllable in poetry is contrary to the spirit of a language. There are slight touches here and there which simply serve as connecting links, and which, in marking the rhythmic flow of sounds, should not be included as belonging to the metre. Most of these are prefixes or affixes, pauses for repose or relaxation, consisting in scarcely noticeable inspirations or expirations, which are necessary to strengthen the voice for the actual metre. The various intonations are generally expressed by the use of the signs for long and short only. As the latter, properly speaking, only represent time or measure, the voice is left to express as best it may and without any guidance whatsoever every other factor composing a language. All I want to do now is to show by the signs for the accent the difference between the English and German rhythmic movement:

′ ˋ ′ ˋ ′ ˋ ′ ˋ
Auf der duftverlornen Grenze
‒ ˘ ‒ ˘ ‒ ˘ ‒ ˘

′ ˋ ′ ˋ ′ ˋ ′ ˋ
Jener Berge tanzen hold
‒ ˘ ‒ ‒ ‒ ˘ ‒

′ ˋ ′ ˋ ′ ˋ ′ ˋ
Abendwolken ihre Taenze
‒ ˘ ‒ ˘ ‒ ˘ ‒ ˘

′ ˋ ′ ˋ ′ ˋ ˋ ˋ
Leicht geschuerzt im Strahlengold.
‒ ˘ ‒ ˘ ‒ ˘ ‒

LENAU.

′ ˋ ′ ˋ ′ ˋ ′ ˋ ′ ˋ
Auf ihrem Grab da steht eine Linde
‒ ˘ ‒ ˘ ‒ ‒ ˘ ˘ ‒ ˘

′ ˋ ′ ˋ ′ ˋ ′ ˋ ′ ˋ
Drin pfeifen die Voegel im Abendwinde;

˗ ˘ ˗ ˘ ˗ ˘ ˗ ˘

╱ ╲ ╱ ╲ ╱ ╲ ╱╲ ╱ ╲ ╱╲ ╱ ╲
Die Winde die wehen so lind und so schaurig,
˗ ˘ ˗ ˘ ˘ ˗ ˘ ˘ ˗ ˘ ˘ ˗

╱ ╲ ╱ ╲ ╱ ╲ ╱╲ ╱ ╲ ╱ ╲ ╱╲ ╱ ╲
Die Voegel die singen so suess und so traurig.
˗ ˘ ˗ ˗ ˘ ˘ ˗ ˘ ˘ ˗

HEINE.

The beginning of every line in this verse might remain unmarked as not belonging to the rhythmic expression proper, and being expressive mainly of an inspiration preceding the expiration which it foreshadows. The beauty of Heine's verse is largely due to the fact that he does not anxiously count time, but lets his voice rise and fall where it is most effective. It will be noticed that there is a greater movement, as expressed by the signs of the rhythm, in Heine's verse than there is in Lenau's, hence the inexpressible charm of his diction. Here is another great poet, or poetess rather, the greatest Germany has produced, also fearless of prescribed forms, but full of charm and power:

╱ ╲ ╱╲ ╱ ╲ ╱╲ ╱╲
O schaurig ists uebers Moor zu gehn,
˗ ˘ ˘ ˗ ˗ ˘ ˗ ˘ ˗

╱ ╲ ╱╲ ╱ ╲ ╱ ╲
Wenn es wimmelt vom Haiderauche,
˗ ˘ ˘ ˗ ˗ ˘ ˘

╱╲ ╱╲ ╱╲ ╱╲ ╱ ╲ ╱ ╲
Sich wie Phantome die Duenste drehn
˗ ˘ ˗ ˗ ˘ ˘ ˗ ˘ ˗

╱ ╲ ╱ ╲ ╱╲ ╱ ╲
Und die Ranke haekelt am Strauche.
˗ ˘ ˗ ˘ ˘ ˘ ˗ ˘

In these last two citations, the dactylus (¯ ˘ ˘) is the prevailing measure, which but strengthens my assertion that in German diction there is a fall after a rise; the former being here more distinctly expressed than in the simple trochaic measure. The fall, the relaxation, being greater, the rise, the vigor in the expression, thereby gains additional strength. What is the consequence of this falling off or gliding down in German diction so well expressed in Lenau's

´ ` ´ ` ´ ` ´ `
"Auf der duftverlornen Grenze"?

It is not a positive line of demarcation, but one which is lost, as it were, "in the soft ether of the evening sky."

Hence the high tide succeeded by the low, the aspiration followed by resignation, the night after the day, death after life, repose after the strife—all this expresses the genius of the German language; and is also expressive of German life and character—its dreaminess, its longing, its desire for the ideal, never to be attained; the abstract, the abstruse; its yearning, its altruism, its transcendentalism, its *Weltschmerz* (the sadness pervading all nature). It is also expressive of its *Begeisterung* (an enthusiasm which upon the slightest provocation takes a man almost off his feet). All these are traits of the German national character.

There is no spiritual bond among all these millions that could possibly produce such sentiments and feelings as its result, differing, as they do, from the feelings of any other nation or people, but that of a language common to all.

To prove that the trochaic measure is the one ordained by nature for German expression, it is but necessary to glance at the characteristic words of the preceding verses:

／ ＼　／ ＼　／ ＼　／ ＼　／ ＼　／ ＼　／ ＼

Wimmelt, Haide, gehen, wehen, drehen, Ranke, haekelt,

／ ＼　／ ＼　／ ＼　／ ＼　／ ＼　／ ＼

Grenze, jener, Berge, Abend, Wolken, Taenze,

／ ＼　／ ＼　／ ＼　／ ＼　／ ＼　／ ＼　／ ＼

strahlen, ihren, eine, Linde, pfeifen, Voegel, Winde,

／ ＼　／ ＼　／ ＼

schaurig, singen, traurig.

The same rhythm, though not so obviously expressed, obtains with the words of one syllable:

／＼　／＼　／＼　／＼　／＼　／＼

Auf, der, Duft, hold, leicht, im, Gold,

／＼　／＼　／＼　／＼　／＼　／＼

Grab, steht, lind, suess, ueber's, Moor.

Now compare with this the strength and vigor of English diction, which runs in the precisely opposite direction:

＼　／＼　／＼　／＼　／

The stag at eve had drunk his fill,
ᵕ　-　ᵕ　-　ᵕ　-　ᵕ　-

＼／　＼／　＼　／＼　／＼　／

Where danced the moon on Monan's rill;
ᵕ　-　ᵕ　-　ᵕ　-　ᵕ　-

＼／　＼／　＼　／＼　／＼　／＼　／

And deep his midnight lair had made,
ᵕ　-　ᵕ　-　ᵕ　-　ᵕ　-

＼　／　＼／　＼　＼／　＼／

In lone Glenartney's hazel shade.
ᵕ　-　ᵕ　-　ᵕ　-　ᵕ　-

SCOTT.

＼　／＼　／　＼　／　＼／　＼／

The day is done, and the darkness
ᵕ　-　ᵕ　-　ᵕ　-　ᵕ

Falls from the wings of night,

As a feather is wafted downward

From an eagle in his flight.

LONGFELLOW.

Oh east is east, and west is west,

And never the two shall meet,

Till earth and sky stand presently,

At God's great judgment seat.

But there is neither east nor west,

Border, nor breed, nor birth,

When two strong men stand face to face,

Though they come from the ends of the earth.

KIPLING.

It is either the iambic (˘ˉ) or the anapest (˘˘ˉ). Of course, these vary to some extent in conformity with the reader's intonation, but the spirit of the language is always from weakness to strength, in place of from strength to weakness, as with the German. It is always the waves approaching the shore and then *breaking* against it, as against the wind *coming up suddenly* and then dying away. This is the reason why a serenade or lullaby in English can never be rendered with the same effect as in German, the English voice rising at the end instead of falling.

Wherever a verse commences with a stress, it must be considered that a fall of the voice or an inspiration has preceded it; this, though unaccompanied by sound, being really the case. I have thus marked the beginning of Longfellow's beautiful lines:

Falls----as----from.

Mr. Lunn, in his *Philosophy of Voice*, has the following:

"How many Englishmen *dare* utter loudly a word beginning with a vowel? If attempted, either it would not be done, or, in spite of the speaker, owing to the weakness of the muscles which draw the cords together [*sic*], an aspirate would precede the vowel."

This is right, as far as his observation is concerned, but he does not seem to know that this very weakness he complains of is really the strength of the English language, the lull before the storm, the concentration before the explosion; and that "thus the idiosyncrasy of our people's speech" is *not* "deadness, weakness, and general feebleness," but, on the contrary, a strength and a virility not surpassed by any other tongue. This finds illustration in Kipling's

˘‾ ˘‾ ˘‾ ˘‾
Oh east is east, etc.

It is but necessary to comprehend the laws which underlie this apparent weakness to turn it to its best account, and to obtain from it the highest results, both for speech and song. As for the "weakness of the muscles which draw the cords together," it will scarcely be necessary for me to make a specific refutation; the premises upon which such assumption is founded being quite untenable, there being quite as much vigor in the *muscles* and *cords* of an Anglo-Saxon as in those of any other nation. Nor, I suppose, will it be necessary to strengthen my assertions by once more quoting the separate words and thus pointing out the iambic, the rise after the fall (˘‾), or the anapest (˘˘‾), the twofold repose and gathering of strength for the final emphasis.

The English language in its Saxon words mainly consists of monosyllables. These, however, as stated, must be looked upon as words of two syllables, a suppressed intonation always preceding their vowel sounds. The majority of such words, as a matter of fact, originally consisted of two syllables, of which the last was dropped when they were adopted by the English. This last syllable, representing the fall of the voice thus disappearing, left the first, which represented its rise, standing unsupported by itself. As the rise of the voice, however, cannot be expressed without the accompaniment of its fall, the latter always *tacitly* accompanies the same, and is expressed in an undertone, *preceding* the rise.

Almost every verb of this class will give evidence of this fact:

Gehen--go, sehen--see, hoeren--hear,

sprechen--speak, kochen--cook, tanzen--dance,

fallen--fall, etc.

Hence, in conformity with the above, these words in the English language should be properly marked thus:

Go, see, hear, speak, cook, dance, etc.

which gives the real intonation thereof.

This applies to all words commencing with a vowel, and explains what Mr. Lunn has designated as a "weakness of the English language":

Art, arm, or, all, eagle, each, old, etc.

Without this half-suppressed fall of the voice, there would be no beauty, no charm, no soul in the English language; in fact, it could not exist. Words of two syllables, however, always have the fall of the voice on the first, its rise on the second, syllable, even where the preponderance of *time* belongs to the first syllable, as in the words

Danced, hazel, etc.

The reader will find these statements sustained by almost every word he may examine into, which will show that the characteristic expression of English diction is that of the iambic measure, which passes from weakness to strength; while that of German diction, as already stated, is that of the trochaic measure, which passes from strength to weakness.

Having shown that German *sentiment* is in accord with the idiomatic expression of the German language, I will now show that *English* sentiment also conforms to *its* idiomatic expression. I must beg the reader, however, not to be over-critical. I am not attempting to furnish comparative sketches of the national character of these peoples in a literary sense, but am entering into these matters for the sole purpose of sustaining the results of my physiological investigations. Nor should these attempts be applied to individual cases, there being exceptions to all rules, but to the national character *in general.* If a person in making investigations of this kind had to constantly fear that he might be treading on some one's sensitive toes, he could never make any headway at all. I am, in fact, perfectly willing to apologize beforehand for any such mishap possibly taking place, as I wish to be perfectly impartial and without bias. I have said this much partly for the reason also that in consequence of some remark, on one occasion, made in my former publication in favor of the English *vs.* the Germans, one critic honored me with the epithet "renegade."

The rising voice succeeding the falling is not a soft and gradual receding, but, on the contrary, it is more like an explosion, a trumpet-blast; the inspiration which had been "stored" being suddenly released. There is no such "storing" in connection with German diction; inspiration and expiration succeeding each other on the spot. With English diction this change may be compared to the break of day after the night; the fray after the repose; resurrection after death; a conflagration and a rebuilding at once on the spot, not only individually, but by an entire community (Boston and Chicago); an outburst after due deliberation; no sentimentality, but a firm resolve for the right; patient submission to a point, then a strike for liberty; the slow accumulation of a fortune and the spontaneous spending thereof; a hot political campaign and a victory or defeat; in either case acquiescence; no vain mourning after the fact; a butterfly of wealth, idleness, and fashion, then perhaps ruin; yet not despair, but a brave conformity to altered circumstances; an energy in the pursuit of business or of war which does not flag until utterly exhausted or success is achieved and a victory is won. All this is due to the reserve force in the character of English-speaking people, which comes to their rescue when circumstances demand it. A world positive and direct, full of energy, restlessness, and activity. A world of, and for, *this* world; whose world to come, even, must have a positive and well-defined character and surroundings:

"Where the walls are made of jasper and the streets are paved with gold."

To what is all this due but to this *bond of language* uniting these millions, and embracing every foreign element, in its children at least? The theme is inexhaustible, but I am limited as to time; yet additional remarks on the same subject will be forthcoming during the further pursuance of these studies.

For song, it appears to me, the words, besides being marked by notes, should also be marked as to rhythm, as this would assist singers in giving them the proper intonation; notes indicating metre, but not rhythm.

Metre and rhythm are produced by two distinctly different processes; metre, or time, being the outcome of a mode of breathing subject to the will, while rhythm is the outcome of an involuntary mode of breathing for a characteristic quality inherent in a nation's language as its idiomatic expression.

Ordinarily, both metre and rhythm are expressed by the same signs (˘¯); this is very misleading.

To express time, or metre, I use the signs for short and long (˘ ˉ). To express rhythm, or the fall and rise of the voice, I use the signs for what is usually called the accent (` ´). If we were to *measure* the exact time, however, consumed in the utterance of syllables, we would find that the falling voice, which is the product of inspiration and belongs to the thorax, requires more time than the rising voice, which is the product of expiration and belongs to the abdomen.

In marking verse, however, the sign for long (ˉ) generally accompanies the short syllable of the rising, and the sign for short (˘) the, as a matter of fact, long syllable of the falling voice. It takes longer to fill a bottle than to pour out its contents; to prepare a dish than to eat it; to walk upstairs than to jump from a window. It takes longer to *prepare* for an utterance than to utter it. It takes longer to inspire than to expire.

In view of the vast foreign element constituting a part of this nation, it would be a matter of interest to know at what period the foreigner ceases to exist as such and the "American" begins; or, in other words, to understand when the evolution takes place which transforms the foreigner into the American. From my point of view it is, above all, a question of language. The political aspect of the case is scarcely to be considered. An unnaturalized Englishman, consequently, after thoroughly "Americanizing" his language, becomes more of an American (no matter whether he himself thinks so or not) than an Irishman who, though naturalized, never ceases to use his native brogue.

These questions, of course, are many-sided. When I speak of nationality, however, I have the *best* specimens of a nation as representatives thereof in view always. A man with a foreign accent does not have the same standing or influence in municipal, state, and national councils as one who speaks a pure English; there is always a *feeling* against him, no matter how able or patriotic he may be, of some foreign influence as a substratum in his composition.

STRESS

I have already stated that the thorax is the seat of the falling, the abdomen that of the rising, voice. This can be tested by a simple experiment, the result of which will be as startling as it is phenomenal. *By simply pressing the stomach, or making the same rigid, you will find that the fact of your doing so will prevent you from uttering any sound belonging to the rising voice, or the stress laid upon a word.*

Take, for instance, the following:

"Oh, say, can you see by the dawn's early light,"

and you will find that, upon pressing the stomach, or making the same rigid, you will not be able to utter the words "say," "see," "dawn's," and "light." This will become more obvious in uttering these words slowly than in doing so rapidly. You will have no difficulty, on the other hand, in uttering the rest of the words, viz.: "Oh," "can you," "by the," "early."

Upon releasing the stomach and bringing a pressure to bear upon the chest, on the other hand, you will have no difficulty in uttering the first words mentioned, those of the rising, while you will beunable to utter the last, those of the falling voice. This rule holds good for all peoples and all languages.

There is this difference, however, as between English and German speech, that, for the former, the falling voice (identical with that of the thorax) *precedes* the rising (identical with that of the abdomen); while for the latter the reverse is the case;— Anglo-Saxons inspiring into the chest and then into the stomach; Germans into the stomach and then into the chest. Germans will have greater difficulty in making this experiment than Anglo-Saxons, as words of the falling voice, as a rule and in all languages, precede those of the rising. Germans, consequently, must *think* of the word of the rising voice, which, as a matter of fact, succeeds the words of the falling, before they can utter the latter. This difficulty is enhanced by the fact that while the rising voice is generally confined to a single word, the falling voice generally embraces several.

Hence the frequency of the use of the anapest (˘˘¯) and the dactylus (¯˘˘), and the relative rarity of the use of the bacchius (˘¯¯) and the antibacchius (¯¯˘); short always representing the falling voice, which embraces more than one word, while long represents the rising voice, which usually embraces but one single word; the definition requiring more words than the thing to be defined. Hence, *for German diction, the "thought" of the word of the rising voice must precede the "utterance" of the words of the falling; while for English diction, the "thoughts" of the words of the falling voice must precede the "utterance" of the word of the rising.*

A German may try and say the following:

"In einem *Thal* bei armen *Hirten,* Erschien mit jedem jungen *Jahr*,"

in such a manner as *not to think* of the words which are italicized before uttering those which immediately precede them, and he will find that he will be unable to pronounce the latter.

An Anglo-Saxon may try and say the following:

"And the star-spangled banner in triumph *doth wave* O'er the land of the free and the home *of the brave*,"

and he will find that in saying "in triumph doth wave," he must think of the words "doth wave" before he will be able to utter the word "triumph." Again, in saying "the home of the brave" he must think of the words "of the brave" before he will be able to utter the word "home."

A German, consequently, must *think* of the principal word before he can utter those which qualify it; an Anglo-Saxon must think of the latter before he can utter the former.

In place of using mechanical pressure, the same results can be obtained by making the respective parts rigid. Regarding this matter of *making parts rigid*, I want to make the following explanation, illustrating the physiological process going on in so doing.

While a part is rendered inactive, placed *hors de combat*, so to say, by the application of mechanical pressure, the same result can also be obtained by making such part rigid. To accomplish this, it is but necessary to positively *think* of such part, to associate your mind with it, which is equal to an act of expiration when it relates to the abdomen, and inspiration when it relates to the thorax. By positively *thinking* of the abdomen, which is equal to an expiration therefrom, you will be unable to utter the stress or *rise* of the voice, which is the product of an expiration from the stomach; by positively thinking of the thorax, which is equal to an inspiration into the same, you will be unable to utter the *fall* of the voice, which is the product of an inspiration into the chest. The reason is obvious: *We cannot utter sound in the same direction in which we breathe; sound and respiration always following opposite directions.*

For the purpose of making satisfactory experiments in this respect, as, in fact, in every other respect in connection with these investigations, it is necessary that inspiration or expiration, as the case may be, should be *continuous*, that is, that either the one or the other should be persisted in until a result is obtained; namely, until an apparent increase or decrease in the size of the part of the body under consideration, or an inflation or depletion of the same, will be perceptible. Though it may be difficult at first, a person will soon learn to distinguish between an increase or a swelling of a part, which means inspiration into the same, and a decrease or a shrinking or diminution thereof, which means expiration from the same.

PHYSIOLOGY OF THE VOICE IN RELATION TO WORDS

In the further pursuance of the questions heretofore under consideration, I shall now enter upon a theme of a still more subtle nature. The question of metre, rhythm, accent, etc., is one which is involved in much mystery; nor can I find that many persons entertain precisely the same ideas as being expressed by these terms.

Accepting as a fundamental principle the fact that our various spiritual conditions are based upon our ability to extract the necessary inspiration therefor from the air, which bears the same relation to our spiritual existence that the earth does to that of our body (in furnishing it with such elements as it requires for its maintenance), I contend that we breathe for speech in as many different modes as there are parts or elements in its composition. This proposition does not necessarily conflict with the fact that we also draw elements from the air, as analytical chemistry has proven, which serve for the construction of matter; such elements, however, instead of being strictly material, as they have every appearance of being, are, in reality, the spiritual complements of the matter they help to form; matter and spirit going hand in hand in our entire composition.

In reading poetry, or giving expression to the same in song (I repeat), we do so in a fourfold manner:

First: as to metre or time (the "measure" of time).

Second: as to the rhythm or the music pervading the voice, produced by its rise and fall, also called cadence, or the idiomatic expression of a language.

Third: as to accent.

Fourth: as to emphasis.

The *metre* is produced by an artistic mode of breathing (in addition to our ordinary and permanent mode), marked by regular repetitions of a given order of inspirations and expirations which can be "measured" as to the time consumed in their enunciation, and are therefore, not incorrectly, called "feet."

The metre is a product or outcome of the *will*, a force which presides over material-spiritual issues. It changes with our inclinations and moods, and is expressive thereof. We can pass from one metre to another at will, as the occasion may require. It is the *material* part of speech, as we can measure it and account for it as to time in space, supposing time to be incorporated. The metre expressive of joy, for instance, being quick, that of sorrow slow; the former, if incorporated, would take up less space than the latter, in the same proportion as it consumes less time in being uttered.

The *rhythm* is that characteristic quality which distinguishes one language from another, the basis upon which it is built and around which all its elementary words cluster; its fundamental principle, its idiomatic expression, the music pervading its every syllable; the inflection, the rise and fall, the cadence of the voice; the spirit of a language, which is permanent and unchangeable.

The rhythm is an outcome of the *mind*; an influence which presides over *spiritual-material* issues. As *harmony is the first law of nature*, so is that harmony which pervades our native tongue the law upon which our individual and national characteristic expressions and actions are based. We exercise it intuitively. It is innate in, and unalterably connected with, our native tongue. It cannot be eliminated therefrom, or put into it by a foreigner, except when acquired in childhood, or by the study of such principles as I have attempted to lay down in this book. It is inborn in every language as its spirit, and is as enduring as that language itself. It is not subject to change by the dictates of the will.

The *accent* represents that element which distinguishes between the character and meaning of words, and has no reference to parts thereof or their relation to other words; the same word being pronounced in as many different ways and with as many different *accents* as it denotes different senses or meanings; while *different words, embodying the same idea, are uttered with precisely the same accent.*

The accent or intonation is an outcome of the *soul*; an influence which dominates over our spiritual nature and over *spiritual issues.* "The rose by any other name would smell as sweet." It is equally true that any other name given to the rose would be pronounced by the same indefinable intonation as its present name, with that same embodiment of the mystery of the soul signifying the flower called "a rose." The *word* "rose," which is the same, or nearly the same, in so many different languages, though possessing the same *spiritual* elements in them all, varies as to measure and rhythm in every one of them.

If the influence of the soul, embodying an idea in a word, through the intonation we give it, were not the same for *all* languages, it would not be possible to translate poetry, and retain, to some extent at least, that which is commonly called "the rhythm" of the original; nor would it be possible to sing a song in another language, and retain, even approximately, the spiritual elements of the original. We would not be impressed with it, would not be *thrilled* by it.

The intonation of a word, expressive of the soul in the embodiment of an idea, is a bond which unites all humanity; not alone the human souls of any special day and generation, but of all days and all generations. But for the fact that the Greek soul is in us to-day, that the native intonation of *their* words is native with us and with *all* mankind, their *dead* tongue would be *absolutely* dead for us. We could find no meaning in it, no beauty, no spirit, no soul. Think of the melody pervading the soul of Homer and emanating from *his* lyre still living and finding an echo in *our* souls! Think of the harmony pervading the soul of Schiller or Tennyson continuing to live, and pervading the souls of the latest generations! Nor could Luther's famous translation of the Bible or its beautiful English version ever have been produced, and after production have made the same impression on the mind, or been read with the same expression of the voice, as the words of this same Bible made upon the minds, and were expressed by the voice, of its original composers, but for the fact *that words of the same meaning, in every language* (aside from metre and rhythm), *are pronounced precisely the same*. It is this universal comprehension of their beauty which gives immortality to the strains of great singers, whether they appear in their original form or are translated (that is, if well translated) into foreign languages, or are set to music and sung either in the one or the other.

If the performances of creating original compositions and their translations were of a mere mechanical order, or were explainable from a mechanical standpoint, no such soul effects could ever be produced. The word, as such, is a *mechanical* contrivance; but its intonation is of the soul, being an emanation of the idea it represents. If our ears were so schooled that by *their "intonation"* we could comprehend the *meaning of words*, we could understand every language upon simply hearing it spoken.

The people of all nations, through their eyesight, form the same conception of an object; the same being impressed upon all minds in the same manner. When a picture thus impressed upon the mind (brain) is reproduced by, or is translated into, vocal utterance, it continues to remain the same with all people. This does not refer to impressions made by material objects alone, but extends to immaterial subjects as well. Hence, knowing the meaning of a word in one language, we can at once conjure up the idea it represents in all languages.

The sight, however, not only impresses our minds through the eye with a given picture, but, as there is a correlation existing between all our faculties, it also impresses the voice with a given inflection, expressive of such impression upon the mind, and of no other impression; any given sight or mental conception of any kind always producing an inflection of the voice corresponding therewith. The vocal expression of an idea might thus be called an *audible* "photographic" reproduction of the impression made by the original object upon the eyesight, and, respectively, upon the brain, or it might be called a phonographic reproduction thereof, supposing that the picture of an object could be impressed upon the wax and could thus become audible. How such a reproduction may be made from an *immaterial* subject would be more difficult to comprehend. Of the fact, however, that an impression from abstract subjects *is* made, and that an audible expression of such impression is produced through the voice, and that this is the case with all people alike, I expect to furnish positive proof in a future publication. The fact of our not being accustomed to distinguish in this manner between various expressions through inflections of the voice is no proof that they do not exist.

The soul impresses every word with a seal of its own, characteristic of the idea it embodies, there being as many accents or inflections of the voice as there are *separate ideas*, or, rather, *groups of ideas*. I beg leave to copy the following from the *Saturday Evening Post* of April 8, 1899:

"Mr. Kipling recently told an interviewer: 'We write, it is true, in letters of the alphabet; but, psychologically regarded, every printed page is a picture book; every word, concrete or abstract, is a picture. The picture itself may never come to the reader's consciousness, but deep down below, in the unconscious realms, the picture works and influences us.'"

The accent is not subject to the will any more than the rhythm. The will can do *this*, however: it can give greater weight, force, and expression, and a wider scope, to the correlated forces of metre, rhythm, and accent, through the

Emphasis which it infuses into them. Through the emphasis, inlet upon inlet is opened, an additional stream of fresh air is infused into them, flooding the spiritual system. Valve upon valve is then opened to let it out. Hence, emphasis is not an "element" of speech proper, but an amplification, an addition to existing elements, rather, impregnating them with the life of the heart, the feelings, the emotions.

In distinguishing in this manner, as I have in the above, between the will, the mind, and the soul, I consider them parts of a great spiritual system intimately connected with corresponding parts of our physical system, but lay no claim as to the correctness of the *terms* I have used. On the contrary, I feel that they are inadequate, and, at most, a makeshift for more fitting expressions. There is a dearth of expressional terms, and I am doing the best I can with such as are at my disposal.

In the same sense, also, I distinguish between material-spiritual, spiritual-material, and spiritual issues; and consider them the outcome, respectively, of the will, the mind, and the soul.

I wish it were in my power to at once fully explain, as far as I am able to offer any explanation at all, how it is *mechanically* possible to express these four elements of metre, rhythm, accent, and emphasis (so widely differing from each other) at one and the same time, by four different modes of breathing, carried on simultaneously, in addition to our regular mode of breathing. The *perfection* of elocution and of singing is to carry on all these various processes simultaneously in as perfect a manner as the subject and the occasion may demand.

I can explain the preceding, in part at least, as follows:

Verse is generally marked by the signs of long and short. While they denote time or metre in the first instance, they are also used to mark what is called "rhythm." Yet, while metre and rhythm are *apparently* of the same order, they are, as a matter of fact, invariably of an inverse order.

We cannot produce two distinctly different expressions while breathing in one and the same direction. While we breathe for metre in one direction, we breathe for rhythm in the opposite direction.

Regarding that mode of breathing expressive of the soul, and pertaining to words in conformity with their *meaning*, and which, in the absence of any more significant word, I have called the "accent," it is of an altogether different order and does not conflict with these other modes of breathing.

Having stated that rhythm and accent are involuntary productions, and that metre alone is subject to the will, we must look to the metre, measure, or time for our guide in our artistic vocal performances. To this, emphasis must be added, as being likewise subject to the will.

As every language has its own time, or tempo, and cannot be properly produced except in conformity therewith, it appears to me that it should be the first aim of vocal science *to ascertain the exact nature of such tempo* for every separate language. *When the correct time is kept, all other component parts of speech fall into line correctly and involuntarily.* Just what the proportionate tempo is for English as against German vocal utterance, I am unable to say, but it is much quicker for the latter than it is for the former.

There is a duality existing between metre and rhythm: the former is voluntary, the latter involuntary. Thus, also, is there a duality between emphasis and accent, of which the former is voluntary, the latter involuntary. Every voluntary factor, not only in vocal utterance, but every voluntary factor in any artistic performance of whatsoever nature, being sustained by an involuntary counter-factor; the same as voluntary and involuntary muscles complement and sustain each other.

Not only every artistic performance, but I dare say *every* act or action of any kind, is of a dual nature. Every separate duality, again, being sustained by a counter-duality, every performance is sustained by four different factors.

When an act is of a material nature and belongs to the hemisphere of the abdomen, it is sustained by four counter-factors belonging to the thorax. When it is of an immaterial nature and belongs to the hemisphere of the thorax, it is sustained by four counter-factors having their seat in the abdomen. Thus every act or action consists of eight movements, or an *octave* of movements.

SIGNIFICANCE OF THE WORD "SCHOOL" IN CONNECTION WITH THE ART OF SINGING

Having established the fact that the rhythmic movements for English and German vocal expression are directly opposed to each other, the one being represented by the iambic, the other by the trochaic measure, there is still a wide field open for investigation as to the idiomatic expression of other languages. This it should not be difficult to determine; personally, I cannot devote the necessary time to this subject even as far as I might be able to do so in connection with other languages of which I have some knowledge. The differences in other tongues, of course, must be embodied in either of the two measures named, as these embrace all others. Whatever may constitute a nation's idiomatic expression must spring from a variation of either of these. While the precedence is given to the abdomen in some and to the thorax in others, the point of gravitation, which according to its location calls for the special manner in which we inspire into and expire from either the one or the other, establishes such variation in the idiomatic expression of *all* tongues.

All that is said about an Italian, a German, or any other "school" (with the exception, perhaps, of what may constitute the difference between what is called "the *old* and the *new* Italian school," and which covers issues of a nature foreign to these investigations) has its proper significance right here: There is no "school" in the sense in which this word is ordinarily used. There are nations and there are languages belonging to such nations. Each nation's language is that nation's "school," and no one nation can go to school with any other nation.

Peasants and the mass of the people generally in Italy, France, Germany, etc., do not visit academies to study vocal art, yet their mode of expression is precisely the same as that of the best vocal artists of these respective countries. I do not mean to say, of course, that the raw material their voices is made up of is as rarefied and artistically trained, but that the composition, the fundamental element thereof, is of precisely the same order as that of their most finished artists. This raw material, on the other hand, in every instance, varies from that of people belonging to every other nation.

The best thing, therefore, to be done, to bring such vocal material as nature has endowed one with up to its greatest perfection, is to have it "schooled" by artists belonging to one's own nation. There may be a time coming, and the same may not be far distant, when methods may be taught by which one may become acquainted with the spirit, and learn the exact mode of the technical expression, of other nations besides one's own. It will then become possible to comprehend these foreign methods and to profit by comprehending them. As long as the principles upon which they are based, however, are not understood, any attempt at singing according to the same will be futile as an accomplishment or an art, and *hurtful* to the voice of the person making the attempt.

Such person will only injure his or her own natural mode of expression, without acquiring the foreign mode.

The idea of learning a certain mode of expression, the Italian, for instance, for singing, and applying it to *all* tongues, is futile and contrary to all reason. We might, with as much show of reason, say that by learning to pronounce one foreign tongue we may apply that knowledge to the pronunciation of every other foreign tongue.

The true state of affairs, and the only one to follow, is, and always will be, this: First, and above all, learn to use your own tongue thoroughly, for *all* purposes of vocal expression. Then learn the use of other tongues for vocal expression in those other tongues only. You cannot apply the technical mode of Italian expression to English vocal utterance any more than you can apply the technical mode of English expression to Italian vocal utterance. An attempt at so doing is quite as preposterous in the one case as it is in the other.

Besides, for the purpose of singing in his own tongue, an Anglo-Saxon does not and should not want to acquire any other mode, as he is by nature in possession of one of the *best* modes of expression. There is none intrinsically purer, none possessed of more vigor or power of expression. There are those with greater softness combined with purity, but lacking strength, as the Italian; and those with more soulfulness combined with strength, but lacking purity, as the German. This native element of purity allied to strength in the Anglo-Saxon, more especially in the English-American, mode of expression is primarily the cause of the high position in the artistic world of the American singer. I ascribe the superiority of the "American" mode of expression over the "English," when untrammelled as in song, in part to the greater personal liberty, the greater want of conventionality, the vast extent of our territory, and our almost constantly clear and unclouded sky; all these being conditions that assist the free exercise of one's natural endowments. To reach the best results in the art of singing, the body as well as the soul must be, as far as possible, untrammelled in any direction. While the idiomatic expression of the English language here and abroad is the same, the social restraint and the conservatism of the English as a nation act against thebest outcome of their gift of song, which demands for its best expression freedom from conventionality or any other constraint.

Each nation is at its best in its own tongue. Our orators are equal to any there are in the world. They do not speak according to the Italian, the German, or any other school. If they did, they would utterly fail and make themselves ridiculous. Why do people, then, want to "speak" in this more expansive and soulful manner, called "singing," in these foreign modes? I know the answer will be that singing and speaking are things quite apart, having no affinity in their mode of production. I shall show, as I have already partly shown, that they are of precisely the *same order*, though different phases of that order; that they cannot be separated; in so far as the elements which belong to speech also belong to song, and those which belong to song also belong to speech; but that they are used in an inverse order in the former as well as in the latter.

Listen to a person breathing just before falling asleep, in a slow, rhythmical order; material objects retire into the background and assume a semi-spiritual shape. This is a similar condition to the one we are in and in which we breathe during the production of song. [By the by, sleep can be induced by thinking of a song, that is, by mentally singing it]. No two nations, however, breathe just alike in that condition, any more than they do during their waking moments; the mode of breathing during sleep being a reversion always of the one which obtains during our waking moments. Our mode of breathing, however, *always* determines our mode of vocal utterance. We can reverse our voice, as we do in whispering, but it is always the same voice, as a garment is the same when we turn it inside out.

Do you know, by the way, that the English whispering voice is the German speaking, and the German whispering the English speaking voice? Try it, and you will find it so. Go on whispering; that is, continue to use your voice in the *same* mechanical manner, but instead of for whispering, use it for speaking aloud, and you will have the exact mode of the other tongue. An Anglo-Saxon, in so doing, will be able to speak German aloud, but not English; a German will be able to speak English, but not German.

Thinking and speaking are of one and the same order. Thought makes the impression of which speech is the expression. If this were not the case, it would not be possible to pass from thinking to speaking or from speaking to thinking at once, and without an effort. To produce English speech, we must think English in a material way, that is, anteriorly, and in so doing produce an instrument from which English material or speech sounds emanate. To produce English song, we must think English in a spiritual way, that is, posteriorly, and in so doing produce an instrument from which English spiritual or song sounds emanate. We cannot think English in either of these two ways and produce German or Italian sounds for speech or song; nor can we produce the latter sounds in any other manner than by *thinking*, either materially or spiritually, in these languages, and in the proper idiomatic manner inherent therein.

How can an English-speaking person, physically and spiritually formed for English expression, and for no other expression, produce proper Italian sounds? She will think Italian in an English way; and, while singing Italian words, produce them with an English expression. That is not singing Italian, however, but English. Is it likely that she will succeed in acquiring the Italian mode of expression while her teacher himself is ignorant of just what that mode consists in, and in what it differs from the native mode of vocal expression of his scholar? You might as well attempt to produce on a violin the sounds of a violoncello or some other instrument.

To illustrate the power of the natural voice, it will but be necessary to call attention to what occurs in almost any concert wherein one of America's own daughters, now "*prima donna assoluta*," is the main performer. She sings a grand aria, the work of an Italian master, highly artistically and perfectly rendered. Musicians are delighted; the public applauds. She reënters, and now the *donna*, changed to a simple American, sings one of England's or America's own songs. The audience, which before had been languidly listening, at the first notes of this song is stirred, electrified, and now listens intently. When she ceases to sing, there is a storm of applause, as to almost shake the house. Where the artistic sense alone had been engaged before, the hearts and the souls of her hearers have now been touched. Yet I have seen the eccentric Von Buelow deliberately take out his handkerchief after such a demonstration and wipe the "desecration" of the "ditty" from the keys of the piano which had accompanied the song, before he deigned to dignify it with one of his "classic" renderings. No doubt he had much contempt for it all: the song, the singer, and the public. The treasures of that "ditty," however, were of an order similar to those hidden within the breast of every one composing that audience. The pearls, floating through the room from the lips of one of its own daughters, had, with a sympathetic touch, stirred it to its very depths, while the foreign "aria" had left it comparatively cold. Supposing an *Italian* singer were to sing an English "aria" in the English language to an Italian audience, and, after that, were to produce one of her own simple Italian songs, would not the effect be the same? Would Italians, in fact, care to listen to her English interpretation, no matter how artistically rendered?

It is an entirely different thing, however, for German or Italian singers to come here and sing their own songs in their own native tongue. Though foreign, the production is genuine. They sing what belongs to them, that in which they live, breathe; they sing their own soul. Such a performance we can comprehend and appreciate, even as we view a foreigner with interest, and honor him for that which is great and good in him, and for which he is distinguished. We can soon *feel* what is genuine and also that which is not; the former being nature's own production, the latter imitated, forced—unnatural. Italians do not sing English or German songs; why should Germans and English-speaking people sing Italian and French songs, to the exclusion, very often, of their own?

It was but recently that I heard a German choral society sing German songs to a delighted American audience. Then came something weird, strange; it was German, yet the words were not German. Looking at the programme, it turned out to be the famous plantation song, "'Way down upon the Suwannee River." The audience looked bewildered; there was no applause, though, judging by the attitude of the singers, they had expected to make this the grand hit of the evening.

The last performance of the great festival of the United German singers in Philadelphia, in 1897, was the production of the "Star-Spangled Banner." Everything in the appearance of the singers showed that this finale was to be the crowning act of the entire festival. All the singers, male and female, participated, and "Old Glory" was waved in the air during the performance. But, as I had feared, it was a complete failure. Instead of the vast audience spontaneously rising to its feet and being carried away by enthusiasm, it remained cold and indifferent, and there was no applause commensurate with what it would have been had the performers sung the words with the true ring in them and the true English accent. The same thing would happen if the "Marseillaise" were sung in France, or the "Wacht am Rhein" in Germany, by foreign singing societies, no matter how excellently schooled, and how artistically rendered.

A similar experience was had by Madame Brinkerhoff, who relates the same in *The Vocalist* of December, 1896, as follows:

"To show how language is imbedded in the *timbre* of the voice, I will relate an incident of last season. On the first night of the representation of the 'Scarlet Letter,' by Damrosch, sung by German singers, I was not surprised or in the least displeased at hearing this beautiful opera sung with the German *timbre* of voice; but after listening to a whole act, I heard no German words; I listened in vain for the shaping of their consonants and vowels, although I heard the German sounds or *timbres*. So I asked the lady seated next to me what language the people on the stage were singing. 'German,' she replied. I said: 'But I hear no German words. Will you kindly listen and tell me when you hear German words?' She listened and replied, 'No, I do not hear German words, but I thought before it was German.' She asked me if it was English. We could not decide it until the lights were turned on, and looked at the programme, which read, 'sung in English.'

"This summer I asked a distinguished singer and teacher of Philadelphia in what language the 'Scarlet Letter' was sung in that city. She replied, 'Oh, German, of course.' 'Did you hear it?' I asked. 'Yes, and I enjoyed it very much, and it was sung in German,' she replied. 'It said in English on the programme,'

I said. 'Well, if I was fooled, a great many more were fooled—beside myself, all our party thought so too. What are you going to do about it?' Gounod says: 'I did not like Italian singing; their tones were attacked so differently from the French method of singing that it was unpleasant at first, but I went again and again, for I could not stay away. I enjoyed it so much.'"

This is what Frau Johanna Gadski had to say in an interview printed in *Werner's Magazine*:

"I have never had any lessons in acting. The director of the Choral Opera told me at the outset that it was better to act by feeling when singing than by instruction. If one studies only acting and singing, one is not always natural. That is the reason why one who does not speak German does not understand the German people and their spirit, is not a German, and cannot sing the Wagner rôles. One must have the German spirit. Sometimes you write here in your papers that German singers cannot sing. I think they sing German rôles very well. One must sing, act, and, above everything, feel at the same time, and then one can speak to the heart of the listener."

Singing in a foreign tongue is, and must be, and always will be (until these things are more thoroughly understood), to a large extent, simply mechanical. Until then, the soul-stirring depth (*der Zauber*) of the native composition will always be wanting. The Anglo-Saxon race has been altogether too dependent upon European continental nations for its examples, its support, and its development in *all* branches of art. This has been more particularly the case in regard to music and song. Though German music, for obvious reasons, which give Germans the preponderance on this field of art, ranks first among nations, still there should be among English-speaking nations a greater native development thereof in harmony with the national expression.

Song, above all, must be national; it must be in harmony with the *genius* of a nation to attain its highest development. It is too closely allied to a nation's speech to be separated therefrom without doing violence to both its music and its meaning. The music and the words *must go together*; their union is as indispensable as it is indissoluble. While we have excellent vocal material in this country, it lacks the proper food for its nourishment. There is no want of poetic compositions. No nation has their superior, or has them in greater abundance. We have the words and the singers; but there is a woful lack of a higher class of compositions for singing. The latter are not at all commensurate with the abundance and the superiority of the talent that is awaiting their appearance.

With compositions on a par with its vocal talent, this nation might rank first among nations in the art of singing. It must stand on its own footing. It must sing its own songs and must be taught by its own teachers. This dictum may provoke indignation in "foreign" vocal teachers. Though I regret the possible consequences to them, this cannot be helped. Science is synonymous with knowledge, and knowledge with truth, and "the truth must be told if the heavens should fall."

BREATHING

All of the preceding, in a manner, may be said to be a preliminary argument for the great truth I claim to have discovered, namely, that *in the sphere of the trunk of our body the material part of our nature is represented by the hemisphere of the abdomen, its immaterial part by that of the thorax; that in the sphere of the head a similar division obtains, in conformity with which it is also divided into hemispheres representing material and immaterial issues; and that every faculty, and the exercise thereof, have their being in a dual action, in close succession, emanating from these hemispheres.*

The first proposition to be proven was that we breathe through the œsophagus, conjointly with the trachea. If all I have said in the preceding has not already convinced the reader of the truth of this statement, I trust the following experiments will thoroughly convince him thereof. These experiments will also furnish additional proof of the fact that English and German modes of respiration are of an inverse order.

Not the slightest fear need be entertained as to the result of these experiments. I have made the same, and others of a similar nature, over and over again, without being in the least discomfited thereby; and I may add that to the fact of having been entirely divested of fear, I largely owe my success in all these undertakings.

If you are an Anglo-Saxon, and make the muscles of your throat rigid, thereby stopping inspiration through the trachea into the thorax, you will soon experience a decided movement of the abdomen, in conformity with which it will first expand anteriorly, then posteriorly, and again anteriorly. There will now be a pause, after which the abdomen will be first expanded posteriorly, then anteriorly, and again posteriorly. This is as far as you can go; you will be compelled to release your hold on your throat after these six movements; the thorax meanwhile remaining passive.

Upon next making the muscles of the back of your neck rigid, equal to those of the œsophagus, the latter being thereby closed to respiration, you will soon experience a decided movement of the thorax, by which it will be first expanded posteriorly, then anteriorly, and again posteriorly. There will now be a pause, after which the thorax will be first expanded anteriorly, then posteriorly, and again anteriorly.

These twelve movements constitute one act of respiration during which inspiration and expiration for thorax and abdomen equalize each other. The first three movements of the abdomen, consisting of an inspiration, an expiration, and an inspiration, constitute what is commonly called an inspiration; the second three movements of the abdomen, consisting of an expiration, an inspiration, and an expiration, constitute what is commonly called an expiration. Of the six movements of the thorax succeeding these, the first three, consisting of an inspiration, an expiration, and an inspiration, are equal to an inspiration; the last three, consisting of an expiration, an inspiration, and an expiration, are equal to an expiration. We thus have four complete respirations, two of which, equal to an inspiration and an expiration, belong to the abdomen; and two, likewise equal to an inspiration and an expiration, belong to the thorax.

Inasmuch as each of these four respirations is composed of three separate movements, one complete respiration consists of twelve separate movements of the respiratory organs. This relates to our ordinary mode of breathing. For vocal utterance, more especially the utterance of a vocal sound, these four respirations are first made for the impression, and are then, in an inverse order, repeated for the expression. This gives us eight movements, or an *octave* of movements, for each vocal sound; these eight movements, as a matter of fact, consisting of twenty-four separate movements of the respiratory organs. These movements, which in our experiment were of relatively long duration, during our ordinary mode of breathing follow upon one another very rapidly; thorax and abdomen, which during our experiment were restrained, ordinarily and when unrestrained, acting and reacting upon one another in quick succession.

The preceding experiment gives us the following result:

ABDOMEN

Movement	1.	Anterior, inspiration.	
"	2.	Posterior, expiration.	*Inspiration.*
"	3.	Anterior, inspiration.	
"	4.	Posterior, expiration.	
"	5.	Anterior, inspiration.	*Expiration.*
"	6.	Posterior, expiration.	

THORAX

Movement	1.	Posterior, inspiration.	
"	2.	Anterior, expiration.	*Inspiration.*
"	3.	Posterior, inspiration.	
"	4.	Anterior, expiration.	
"	5.	Posterior, inspiration.	*Expiration.*
"	6.	Anterior, expiration.	

All of the preceding has reference to the Anglo-Saxon mode of breathing.

Germans, under the same circumstances, will make movements of an inverse order.

The first movement of the abdomen will be posterior, the next anterior, the third posterior, which will be succeeded by anterior, posterior, and anterior ones; while the movements of the thorax will be anterior, posterior, and anterior, succeeded by posterior, anterior, and posterior ones. This shows that *with Germans, expiration antecedes inspiration,* while *with Anglo-Saxons, inspiration antecedes expiration.*

In our experiment, with Anglo-Saxons, *inspiration* took place in the abdomen by two movements anteriorly to one posteriorly, and in the thorax by two movements posteriorly to one anteriorly; while *expiration* took place by two movements of the abdomen posteriorly to one anteriorly, and in the thorax by two movements anteriorly to one posteriorly, as per this schedule:

ANGLO-SAXON

Abdomen

1. Inspiration, Ant., post., ant.

2. Expiration, Post., ant., post.

Thorax

3. Inspiration, Post., ant., post.

4. Expiration, Ant., post., ant.

In the case of a German, it would have been more proper, for our experiment, to have *first* closed the muscles to the œsophagus, and then those to the trachea, as Germans first breathe into the œsophagus and then into the thorax. Had this been done, the result would have been inverse to that of our experiment, as follows: The first movement of the thorax would have been one of inspiration, the same as the first movement of the abdomen; and the second movement of the thorax would have been one of expiration, the same as the second movement of the abdomen, thus:

GERMAN   **Thorax**

1. Inspiration, Ant., post., ant.

2. Expiration,   Post., ant., post.

Abdomen

3. Inspiration, Post., ant., post.

4. Expiration,   Ant., post., ant.

This shows that the movements of the abdomen are the reverse of those of the thorax:

With *Anglo-Saxons*, in such a manner that, while for the abdomen *inspiration* takes place anteriorly, it takes place for the thorax posteriorly; and that, while for the abdomen *expiration* takes place posteriorly, it takes place for the thorax anteriorly;

With *Germans*, in such a manner that, while for the thorax *inspiration* takes place anteriorly, it takes place for the abdomen posteriorly; and that, while for the thorax *expiration* takes place posteriorly, it takes place for the abdomen anteriorly.

These various modes of breathing find an illustration in the following:

Anglo-Saxons, while carrying a burden (for which purpose it is necessary to hold the breath or to economize the same as much as possible), inspire into the abdomen anteriorly and the chest posteriorly, and in so doing expand the same accordingly; while Germans, under the same circumstances, breathe into and expand the abdomen posteriorly and the chest anteriorly. The action of the former tending away from the diaphragm, that of the latter tending towards it, exercise an influence on the spinal column which causes Anglo-Saxons while carrying a burden to assume an erect, Germans a stooping position. This has already been illustrated by calling attention to the difference between the position of the Greek and Gothic caryatides, the former representing the Anglo-Saxon, the latter the German mode of breathing. The order for German soldiers, "Brust heraus, Bauch herein"! ("Breast out, belly in"), for Anglo-Saxons should be, "Breast in, belly out"! The former gives German soldiers that stiff appearance, tending towards the diaphragm, of which Heine has said:

"Als haetten sie verschluckt den Stock, Womit man sie einst gepruegelt."

("As if the stick they'd swallowed With which they once were walloped.")

The fact that inspiration always consists in an inspiration, an expiration, and an inspiration, while expiration consists in an expiration, an inspiration, and an expiration, is one of the most interesting observations I have made in connection with these studies.

These facts may be generalized in saying: There is no action connected with life which consists of a single movement in any one single direction; every action, of whatsoever nature, if it is outgoing, consisting of an outgoing, ingoing, and outgoing movement; if it is ingoing, of an ingoing, outgoing, and ingoing movement; every superior movement consisting of a superior, an inferior, and a superior; every inferior, of an inferior, a superior, and an inferior one; every left movement, of one to the left, to the right, and to the left; every right movement, of one to the right, to the left, and to the right; the last movement *only* being visible and accompanying action.

While our experiment is representative of the general principles underlying our mode of breathing, the act of breathing, proper, is subject to many variations. During their waking moments, or for conversation, with Anglo-Saxons respiration takes place by thorax and abdomen changing off, alternately, while with Germans they succeed one another in the same manner as they did in our experiment, commencing, however, with the thorax instead of with the abdomen, and with expiration instead of with inspiration, as follows:

ANGLO-
SAXON

1. Insp.   Thorax—post., ant., post.

2.   "   Abd.—ant., post., ant.

3. Exp.   Abd.—post., ant., post.

4.   "   Thorax—ant., post., ant.

GERMAN.

1. Exp.   Thorax—post., ant., post.

2. Insp.   "   —ant., post., ant.

3. Exp.   Abd.—ant., post., ant.

4. Insp.   "   —post., ant., post.

This shows an indirect movement for Anglo-Saxon, a direct movement for German respiration. Hence, English enunciation is necessarily slow, German relatively quick. It also shows that the reserve force with Anglo-Saxons is held before it is expended; with Germans it is expended almost as fast as it is engendered.

As there is an apparent discrepancy between the last schedule and the previous one showing Anglo-Saxon mode of inspiration, I want to remind the reader that our "experiment" was made mainly to set forth the fact that we breathe through the œsophagus conjointly with breathing through the trachea; but it was not intended to show our regular mode of breathing.

Though Germans and Anglo-Saxons breathe in opposite directions, still there is an affinity between them in so far as they breathe *along the same plane.* Peoples who speak any of the Latin tongues, on the other hand, breathe along a different plane, and so do Slavonic, Mongolian, and other races. Anglo-Saxons and Germans, therefore, though opposed to one another in one sense, are affiliated in another; and both may be, therefore, as they often are, said to belong to the Teutonic race, together with other peoples along the borders of the North and Baltic Seas. In a similar manner, no doubt, other races possess their similitudes and dissimilarities.

It should scarcely require any further proof on my part after this and all I have previously said to show that, if any of the peoples now speaking Latin tongues were in place thereof to speak English or German, they would, in the course of time, cease to be Frenchmen, Spaniards, or Italians, as the case might be, and would become Anglo-Saxons or Germans; or that, if any of the Slavonic races or peoples would do the same, the same result would eventually ensue; and also that, if Anglo-Saxon or German peoples were to speak Latin or Slavonic tongues in place of their own, they would eventually cease to be Anglo-Saxons or Germans, and would become the people whose tongue they were speaking; always provided, of course, that such tongues were to be spoken *idiomatically* correctly. Should any one still doubt that language is the mainspring formulating peoples and nations in all that essentially belongs to them and distinguishes them as such, I confidently believe that that which I shall still further have to say on this subject will eventually convince even the most obdurate of the correctness of these assertions.

The preceding schedules both for English-and German-speaking peoples show their mode of breathing during their waking moments and for the purpose of conversation. During sleep and for the demands of the singing voice, however, thorax and abdomen interchange with one another in so harmonious a manner that their inspirations and expirations appear as one respective inspiration and expiration.

The following schedules will show the relation of metre and rhythm to breathing.

Inspiration being of longer duration than expiration, I have in the following signified the former by the sign for long (¯), the latter by that for short (˘); while for the rise of the voice I have used the sign for acute (´), and for its fall that for grave (`).

Anglo-Saxon Abdomen	Thorax
1. Inspiration, ˋ ´ ˋ / ˘ ¯	3. Inspiration, ˋ ´ ˋ / ˘ ¯
2. Expiration, ´ ˋ ´ / ¯ ˘ ˘	4. Expiration, ´ ˋ ´ / ¯ ˘ ˘

An experiment may be made by an Anglo-Saxon adopting the German mode of breathing and then attempting to speak English, or by a German adopting the Anglo-Saxon mode of breathing and then attempting to speak German, which neither will succeed in doing.

In making the experiments just now under consideration, it will *not* be necessary, after closing the muscles of the trachea or the œsophagus for the first six movements, to continue doing so, as the next six movements will ensue involuntarily. There may be several repetitions of these twelve movements involuntarily or automatically following after that; any special mode of breathing once assumed being apt to continue indefinitely until another mode is inaugurated.

The same experiments may also be made by making *abdomen and thorax* alternately *rigid*, or producing a state of rigidity through mechanical pressure, in place of producing it with the muscles of the œsophagus and the trachea. As this may appear simpler and "less dangerous," there should be nothing to hinder any one from making these experiments. The movements will not be as *pronounced*, however, in the latter instance as they are in producing a *direct* closure of the trachea and the œsophagus.

There is a fourth mode of producing the same results, namely, through the simple act of *continuously* "thinking" of any particular part. We may thus bring about a closure of the muscles of the trachea or œsophagus, of thorax or abdomen, etc.; thought, which *precedes* motion for vocal utterance, *always*, as cause to effect, being the final arbiter in all matters of respiration, unless the latter is of an involuntary and simply functional character. While the act of breathing for life pursues its even tenor, breathing for vocal utterance, though of the same *order*, is subject to innumerable changes in conformity with the sound, syllable, or word intended to be produced.

I am aware that there may be *apparent* incongruities in some of the preceding, and I presume there always will be. We can see things only from our limited standpoint. I have undertaken to solve matters supposed to be superhuman, or "of God," and hence *perfect* in their way, in a human, and therefore imperfect, manner. Our limitations naturally extending to our power of observation, the duality of our nature in matters of this kind does not permit us—I might say, forbids us—arriving at *final* conclusions. We can go as far as our understanding permits us to go—beyond that, we may at most indulge in speculation. I have limited myself to my limits, to what I could prove, and have but rarely indulged in what I could not—in speculation.

NOTE.—Since the above was written Dr. G. E. Brewer, who in conjunction with Dr. F. C. Ard, last month (March, 1899), in New York, successfully performed the very rare operation of laryngectomy, has told me that his patient had already (after a month) commenced to speak again, though as yet only in a monotonous whispering voice. She is doing so in spite of the fact that every vestige of her larynx, which had been in a diseased state, and which the doctor showed me, had been removed. When I told the doctor this mysterious "new" voice was that of the œsophagus and had always existed with his patient, as it exists with every one else, and had always been heard in conjunction with that of the trachea, he was greatly astonished, though naturally incredulous, but said he would investigate.

SONG, SINGERS, AND PHYSIOLOGY

We are incomprehensible and mysterious beings. We do not know whence we come nor whither we go; we do not know what agencies guide and sustain us—our end is a tragic one. While the soles of our feet closely adhere to the ground, our heads are in touch with the most distant stars. We exercise faculties to perfection whose origin and mode of operation are unalterably hidden from our knowledge. We possess gifts and talents which raise us above the plane of our ordinary existence and inspire us with the belief that we are related to the divinity, are part of the divinity. It has ever been man's aim to penetrate this darkness, to learn to comprehend *himself*. The vocation of the singer is one to which this knowledge is indispensable. In the fulness of his organization endowed by nature with a divine gift, the singer's aim and desire is to retain and perfect this gift.

The birds sing their same individual song throughout their career. Man, however, sings the song of his soul; a song as endless and as varied as his thoughts. Song with him is not a gift alone, but its exercise is a study, an art. He must sing *knowingly*; he must ascertain the source of his song and the reason why certain causes produce certain results. Hence the necessity for a science of the voice.

The knowledge of the exercise of our faculties is dependent on the knowledge of life and on that of the spirit, without whose aid no transaction of life of any kind ever takes place. Despairing of his ability to penetrate into the realms of the spirit, aspiring man has ever resorted to that which was next at his command—matter. Hence the effort throughout all of man's history to reach the soul by way of the body. But body and mind, in alliance, have ever succeeded in frustrating these efforts; in keeping the secret of their duality and mutuality intact from the gaze of man. Yet singers are determined to find out *something* in relation to the *voice* at least. Finding that we cannot penetrate into the relation existing between mind and matter, the effort is renewed in the most persistent manner to explain the life and the spirit, whose essence and outcome is the voice, by examining into the relation of matter to matter.

Our professor, having discarded the assistance of life and the spirit, dabbles in matter pure and undefiled. This process our young students are invited to attend. They carry their youth and their talent, their high hopes and aspirations, into the dissecting-room, where the spirit of the voice is supposed to reveal itself among the ghastliest spectacles. If a person of ordinary good sense, but not acquainted with these subjects, were to attend a lecture on the physiology of the voice and then attend a singing-lesson based upon the knowledge thus attained, he would be apt to remark: "Can this performance possibly be meant to be in good faith? Is not this man taking advantage of the credulity of this woman, who is giving him her hard-earned money, but to find before long that she has been beggared, not only in purse, but in voice and spirit as well; that she has not been benefited in any sense, but sadly robbed and betrayed?"

The persistency with which the modern scientist attempts to hammer a voice out of the larynx and surrounding material tissues and other physical agencies is a cardinal sin against the holy "spirit." When he uses this supposed knowledge for coining it into money at the expense of trusting and aspiring singers, he commits a malpractice, for which some day he will have to go to the penitentiary of his own conscience; that is, if he is in possession of any. "Vocal bands, mucous membranes, tissues, ligaments, muscles, hollow spaces, air-pressure,"— these are the factors productive of the voice divine; matter, nought but matter; not a spark of the divine afflatus, not a spark even of life.

Journals devoted to the voice are full of these things. I will quote but a single instance. At the Music Teachers' National Convention, held in New York, in June, 1898, a sensation was created by Dr. Frank E. Miller (see *Werner's Magazine* for August, 1898, page 490) saying:

"In other words, I wish to say that the action of the cavities or hollow spaces is anterior and prior to the action of the vocal bands in production of tone and tone-quality in our organs of speech. *With this novel fact I announce an original discovery.*"

It is such *stuff* as this that these people feed upon and believe in as revelations of great moment. Yet Dr. Miller and his coadjutors might sit before these cavities or hollow spaces till the end of time, looking, observing, probing, measuring, weighing, and determining their relation to the vocal bands and vice versa, and not a vestige of the spirit of the voice would ever make its appearance. The last conundrum of this kind, and it has special reference to my discoveries, is as follows: "May not the disturbance of speech known as stammering or stuttering be mainly a condition caused by the putting out of gear of one air-chamber in its relationship to other air-chambers, whereby the air-pressures during the speech-act are at war with one another, resulting in the well-known manifestations?" (*Werner's Magazine* for September, 1898, page 59). Air-chambers and air-pressures again. I protest against being made *particeps criminis* in any such proceeding.

When we go back to the earliest recorded times and find traces of an attempt at expression by means of crude signs or figures impressed upon the clay, we can see more of the potentiality of a science (or a civilization) arising therefrom than we can from the teachings of the laryngoscopists, who claim that the voice can be evolved from the relations of various forms of matter to one another, without even a trace of the spirit accompanying them.

Not many years since audiences of intelligent persons were invited to watch a dark tent in which two men were so closely tied together (as it was supposed) that they could not possibly move a limb. From this tent noises would arise as of the dragging of chains along the floor, bells ringing, etc., interposed now and then by a chair being flung through the air. All this was done by the "spirits." This was a proceeding not unlike the one now going on in the materialistic school in connection with the spirit of the voice. There is no more likelihood of the latter arising from the dark tent of the matter they are investigating than of a real spirit appearing in that other tent. The performance, besides, is not as amusing, no chairs being flung, etc. The audience is looking on gravely expectant, but all remains forever monotonously, solemnly, ominously, and cadaverously silent and resultless.

The *living* grain of corn a blind hen after much scratching succeeds in digging out from beneath a barn-yard floor bears a closer resemblance to life, and hence to the voice, than the relations a professor of physiology scratches together out of the various parts which he supposes make up the instrument of the voice. These attempts are so contrary to reason and common sense that in any other science their originators would be laughed to scorn for their pains.

The other great issue with physiologists in connection with the voice is that of breathing. Clavicular breathing, costal breathing, diaphragmatic breathing, etc.—these are some of the terms in common use, and the "modes" of breathing commonly practised. Each of these modes is supposed to be practised separately and at the will of the performer. They are praised and recommended or condemned according to the special view of the practitioner. Systems are based on these special modes and schools arise therefrom. What one "school" practises is condemned by another. And how could it be otherwise, *all* being wrong?

Being homogeneous entities, whose wholesome existence is based upon a harmonious coöperation of all parts, we cannot practise breathing from a special part without every other part more or less participating. The act of breathing being our most vital performance, every other part would suffer if it were confined to any special part. Our entire system, therefore, must participate therein; the hemisphere of the abdomen no less than that of the thorax; both hemispheres coöperating with each other and with other streams introduced into our system through the pores and every other opening in the body. For a moment, and for an especial expression, one part may prevail over another; but the true artist will always breathe in such a manner that after such an effort all parts will again harmonize and balance one another. He will have such control over his breathing powers that he can at any time throw the balance of power into one direction; but he will never let any one direction *continue* to prevail over any other.

Every theory heretofore advanced in respect to our mode of breathing, being based upon false premises, is wrong in the abstract, and impossible of practical execution.

If I have expressed myself strongly, it is because I feel strongly the injury which has been wrought by this so-called "science" of the laryngoscopists. It has in thousands of instances hindered the natural development of the voice, and has in many other directions done incalculable harm; while it has in *no* direction ever done any good. It has oppressed the intellect, depressed the spirit, and suppressed the soul of singers. Let me add but this: What would be the use of the most scientifically constructed stove, filled with the most appropriate fuel, if the flame were wanting to set fire to this fuel? Supposing the laryngoscopists to comprehend the intricate construction of the stove (the body), the highly sensitive and complicated apparatus of the fuel (the instrument of the voice)—both of which, however, they are greatly in the dark about—the flame would still be wanting to set fire to this fuel and fill the stove with the holy glow of song. This flame (the life, the spirit) they do not even pretend to be able to furnish. They only give us the stove and the fuel, which remain forever dark, cold, lifeless, inert.

To set myself up in judgment regarding these important issues, or to place my judgment over that of so many eminent persons in the past as well as the present, may appear to be a presumptuous, rash, bold, and almost unwarranted undertaking. It is not my fault, however, that there should be such utter confusion existing in these matters; that no one should have ever succeeded in reducing this chaos to any kind of order; that I am the heir, so to say, to this condition of affairs; the trustee to this inheritance, who is to make use of it to the best advantage of all that are interested.

Nor is it my fault that, not by dint of superior endowments, or any other qualities of a superior order, but simply through the discovery of the dual nature of the voice, I should have obtained an insight into, a mastery over, these matters never before enjoyed by any man. Yet there seems to be a disposition on the part of some persons to throw blame on me for these facts; in place of furthering, to suppress, this knowledge; in place of probing and investigating, to assume that it is simply the outcome of a somewhat more than lively imagination. It appears to me that this is partly done in the interest of the vast literature on these subjects now in existence, which will become obsolete and valueless as soon as the *truth* in matters of the voice has been established.

I dare say this simple fact, "We breathe and speak through the œsophagus in conjunction with breathing and speaking through the trachea," for *real* knowledge, is worth all of the entire literature on the voice, as a science, now in existence.

The science of the voice, as I understand and am trying to explain and establish it, is one not so much of mechanical issues, though they have their share in it, as one in which the spirit, this heretofore unapproachable issue, performs the greatest and most vital part. It is a question of life, and every issue and every agency governing life are involved in it. How vast a science this science of the voice therefore is, can be better imagined than at once fully comprehended. I am far from being able to present it in all its aspects, but shall endeavor, as I have already partly done, to continue to give a general outline of it.

It will take time and patience for any one to acquire this knowledge, but the reward will be more than commensurate. To superficially obtain it from others is not sufficient; one must learn to know it of one's own knowledge. It is an academic study, embracing many sciences. A person must enter into it with his whole being if he wants to get hold of the spirit thereof and be truly benefited thereby. He must identify himself with this knowledge, must become part and parcel thereof, or it must become part and parcel of him. When this is done, true teachers of the voice will arise, for here is a chance for greatness to assert itself. It will be death to all hackneyed knowledge and charlatanism.

When the true knowledge of the production of speech and song for *every* language has been established, when we have a real science of the voice, the teacher comprehending these issues in their entire latitude will be able to teach how to interpret Mozart, Schubert, and Wagner, Rossini and Verdi, Gounod, and every other master in the tongue and the spirit in which he has produced his works.

The genius for execution in the art of singing is with the Anglo-Saxon race, but not for composition, for original conception. It may come, but it is not with it now.

The desire of the singer naturally is to embrace the highest in her or his repertoire. At present it is Wagner. But how can Wagner be rendered without a comprehension of his genius as expressed through his language? The genius of the master and the genius of the language he wrote and composed in cannot be separated. They are soul and body of one and the same entity. Without the comprehension of the genius of the German language, of its idiomatic expression, it is not possible to reproduce what Wagner meant to express by his work. To sing German with an English tongue is an anomaly; it is still English in the real sense of the word, and not German. It is an unnatural proceeding, and therefore injurious to the vocal organs of the singer.

No one would expect a foreigner, for the delectation of a native-born audience, to recite before it poetry in the latter's language, or a native-born person to recite before it in a foreign tongue. In either case such a person would fail. Why, then, song, this sister art and accomplishment?

All these are questions which, though ever so reluctantly, artists will have to face. It complicates their art, but it will also, when understood, make it comparatively easy. Americans will then sing the works of foreign masters with the same perfect ease that they do those of their native composers, and so will persons of every other nationality.

Who will be able to teach a foreign language so well as the natives of each respective country? provided such persons have learned to comprehend the difference between the mode of production of their speech and that of their scholars. In that case only will a German be able to teach an Anglo-Saxon his (the German) language for either speech or song. It will be the same with every other nationality.

The teachers, as a class, are with me. They feel that the efforts of the physiologists to aid them in their vocation are wrong and misleading. They have no faith in the revelation of matter. They know matter is inert, powerless for any purpose without the indwelling of the spirit; that the spirit reigns over and controls *every* manifestation of life; and that the voice in singing is one of the highest manifestations thereof. They know that song comes from the heart and the soul, while it uses the body for its instrument.

I have been told I must build up before tearing down; before destroying the old I must put something better in its place. I think it a praiseworthy undertaking, in itself, to destroy the false and the harmful. Besides, we cannot erect a new building before the old one has been removed.

As for this *new* science, I am doing what I can to put it into shape, to give a visible and tangible form to it as it has developed in my mind. The world has been able to do without it so long, those interested in these matters must have a little patience.

I specially appeal to the *young* to devote themselves to these studies and to thus become the precursors in the application of principles which are destined to revolutionize the vocal science of the world; the old being often too old to get out of lifelong practices, no matter how erroneous. I appeal in like manner to the students of medicine, and to those of every other branch of science, whose aim is the knowledge of man in any of, and all, his relations.

www.ingramcontent.com/pod-product-compliance
Lightning Source LLC
Chambersburg PA
CBHW050210230526
45470CB00001B/319